Geology: A Very Short Introduction

VERY SHORT INTRODUCTIONS are for anyone wanting a stimulating and accessible way into a new subject. They are written by experts, and have been translated into more than 45 different languages.

The series began in 1995, and now covers a wide variety of topics in every discipline. The VSI library currently contains over 550 volumes—a Very Short Introduction to everything from Psychology and Philosophy of Science to American History and Relativity—and continues to grow in every subject area.

Very Short Introductions available now:

ACCOUNTING Christopher Nobes
ADOLESCENCE Peter K. Smith
ADVERTISING Winston Fletcher
AFRICAN AMERICAN RELIGION Eddie S. Glaude Jr
AFRICAN HISTORY John Parker and Richard Rathbone
AFRICAN RELIGIONS Jacob K. Olupona
AGEING Nancy A. Pachana
AGNOSTICISM Robin Le Poidevin
AGRICULTURE Paul Brassley and Richard Soffe
ALEXANDER THE GREAT Hugh Bowden
ALGEBRA Peter M. Higgins
AMERICAN HISTORY Paul S. Boyer
AMERICAN IMMIGRATION David A. Gerber
AMERICAN LEGAL HISTORY G. Edward White
AMERICAN POLITICAL HISTORY Donald Critchlow
AMERICAN POLITICAL PARTIES AND ELECTIONS L. Sandy Maisel
AMERICAN POLITICS Richard M. Valelly
THE AMERICAN PRESIDENCY Charles O. Jones
THE AMERICAN REVOLUTION Robert J. Allison
AMERICAN SLAVERY Heather Andrea Williams
THE AMERICAN WEST Stephen Aron
AMERICAN WOMEN'S HISTORY Susan Ware
ANAESTHESIA Aidan O'Donnell
ANALYTIC PHILOSOPHY Michael Beaney
ANARCHISM Colin Ward
ANCIENT ASSYRIA Karen Radner
ANCIENT EGYPT Ian Shaw
ANCIENT EGYPTIAN ART AND ARCHITECTURE Christina Riggs
ANCIENT GREECE Paul Cartledge
THE ANCIENT NEAR EAST Amanda H. Podany
ANCIENT PHILOSOPHY Julia Annas
ANCIENT WARFARE Harry Sidebottom
ANGELS David Albert Jones
ANGLICANISM Mark Chapman
THE ANGLO-SAXON AGE John Blair
ANIMAL BEHAVIOUR Tristram D. Wyatt
THE ANIMAL KINGDOM Peter Holland
ANIMAL RIGHTS David DeGrazia
THE ANTARCTIC Klaus Dodds
ANTHROPOCENCE Erle C. Ellis
ANTISEMITISM Steven Beller
ANXIETY Daniel Freeman and Jason Freeman
APPLIED MATHEMATICS Alain Goriely
THE APOCRYPHAL GOSPELS Paul Foster
ARCHAEOLOGY Paul Bahn
ARCHITECTURE Andrew Ballantyne
ARISTOCRACY William Doyle
ARISTOTLE Jonathan Barnes
ART HISTORY Dana Arnold

ART THEORY Cynthia Freeland
ASIAN AMERICAN HISTORY Madeline Y. Hsu
ASTROBIOLOGY David C. Catling
ASTROPHYSICS James Binney
ATHEISM Julian Baggini
THE ATMOSPHERE Paul I. Palmer
AUGUSTINE Henry Chadwick
AUSTRALIA Kenneth Morgan
AUTISM Uta Frith
AUTOBIOGRAPHY Laura Marcus
THE AVANT GARDE David Cottington
THE AZTECS Davíd Carrasco
BABYLONIA Trevor Bryce
BACTERIA Sebastian G. B. Amyes
BANKING John Goddard and John O. S. Wilson
BARTHES Jonathan Culler
THE BEATS David Sterritt
BEAUTY Roger Scruton
BEHAVIOURAL ECONOMICS Michelle Baddeley
BESTSELLERS John Sutherland
THE BIBLE John Riches
BIBLICAL ARCHAEOLOGY Eric H. Cline
BIG DATA Dawn E. Holmes
BIOGRAPHY Hermione Lee
BLACK HOLES Katherine Blundell
BLOOD Chris Cooper
THE BLUES Elijah Wald
THE BODY Chris Shilling
THE BOOK OF MORMON Terryl Givens
BORDERS Alexander C. Diener and Joshua Hagen
THE BRAIN Michael O'Shea
BRANDING Robert Jones
THE BRICS Andrew F. Cooper
THE BRITISH CONSTITUTION Martin Loughlin
THE BRITISH EMPIRE Ashley Jackson
BRITISH POLITICS Anthony Wright
BUDDHA Michael Carrithers
BUDDHISM Damien Keown
BUDDHIST ETHICS Damien Keown
BYZANTIUM Peter Sarris
CALVINISM Jon Balserak
CANCER Nicholas James
CAPITALISM James Fulcher
CATHOLICISM Gerald O'Collins
CAUSATION Stephen Mumford and Rani Lill Anjum
THE CELL Terence Allen and Graham Cowling
THE CELTS Barry Cunliffe
CHAOS Leonard Smith
CHEMISTRY Peter Atkins
CHILD PSYCHOLOGY Usha Goswami
CHILDREN'S LITERATURE Kimberley Reynolds
CHINESE LITERATURE Sabina Knight
CHOICE THEORY Michael Allingham
CHRISTIAN ART Beth Williamson
CHRISTIAN ETHICS D. Stephen Long
CHRISTIANITY Linda Woodhead
CIRCADIAN RHYTHMS Russell Foster and Leon Kreitzman
CITIZENSHIP Richard Bellamy
CIVIL ENGINEERING David Muir Wood
CLASSICAL LITERATURE William Allan
CLASSICAL MYTHOLOGY Helen Morales
CLASSICS Mary Beard and John Henderson
CLAUSEWITZ Michael Howard
CLIMATE Mark Maslin
CLIMATE CHANGE Mark Maslin
CLINICAL PSYCHOLOGY Susan Llewelyn and Katie Aafjes-van Doorn
COGNITIVE NEUROSCIENCE Richard Passingham
THE COLD WAR Robert McMahon
COLONIAL AMERICA Alan Taylor
COLONIAL LATIN AMERICAN LITERATURE Rolena Adorno
COMBINATORICS Robin Wilson
COMEDY Matthew Bevis
COMMUNISM Leslie Holmes
COMPARATIVE LITERATURE Ben Hutchinson
COMPLEXITY John H. Holland
THE COMPUTER Darrel Ince
COMPUTER SCIENCE Subrata Dasgupta
CONFUCIANISM Daniel K. Gardner
THE CONQUISTADORS Matthew Restall and Felipe Fernández-Armesto
CONSCIENCE Paul Strohm

CONSCIOUSNESS Susan Blackmore
CONTEMPORARY ART Julian Stallabrass
CONTEMPORARY FICTION Robert Eaglestone
CONTINENTAL PHILOSOPHY Simon Critchley
COPERNICUS Owen Gingerich
CORAL REEFS Charles Sheppard
CORPORATE SOCIAL RESPONSIBILITY Jeremy Moon
CORRUPTION Leslie Holmes
COSMOLOGY Peter Coles
CRIME FICTION Richard Bradford
CRIMINAL JUSTICE Julian V. Roberts
CRIMINOLOGY Tim Newburn
CRITICAL THEORY Stephen Eric Bronner
THE CRUSADES Christopher Tyerman
CRYPTOGRAPHY Fred Piper and Sean Murphy
CRYSTALLOGRAPHY A. M. Glazer
THE CULTURAL REVOLUTION Richard Curt Kraus
DADA AND SURREALISM David Hopkins
DANTE Peter Hainsworth and David Robey
DARWIN Jonathan Howard
THE DEAD SEA SCROLLS Timothy H. Lim
DECOLONIZATION Dane Kennedy
DEMOCRACY Bernard Crick
DEMOGRAPHY Sarah Harper
DEPRESSION Jan Scott and Mary Jane Tacchi
DERRIDA Simon Glendinning
DESCARTES Tom Sorell
DESERTS Nick Middleton
DESIGN John Heskett
DEVELOPMENT Ian Goldin
DEVELOPMENTAL BIOLOGY Lewis Wolpert
THE DEVIL Darren Oldridge
DIASPORA Kevin Kenny
DICTIONARIES Lynda Mugglestone
DINOSAURS David Norman
DIPLOMACY Joseph M. Siracusa
DOCUMENTARY FILM Patricia Aufderheide
DREAMING J. Allan Hobson
DRUGS Les Iversen
DRUIDS Barry Cunliffe
EARLY MUSIC Thomas Forrest Kelly
THE EARTH Martin Redfern
EARTH SYSTEM SCIENCE Tim Lenton
ECONOMICS Partha Dasgupta
EDUCATION Gary Thomas
EGYPTIAN MYTH Geraldine Pinch
EIGHTEENTH-CENTURY BRITAIN Paul Langford
THE ELEMENTS Philip Ball
EMOTION Dylan Evans
EMPIRE Stephen Howe
ENGELS Terrell Carver
ENGINEERING David Blockley
THE ENGLISH LANGUAGE Simon Horobin
ENGLISH LITERATURE Jonathan Bate
THE ENLIGHTENMENT John Robertson
ENTREPRENEURSHIP Paul Westhead and Mike Wright
ENVIRONMENTAL ECONOMICS Stephen Smith
ENVIRONMENTAL LAW Elizabeth Fisher
ENVIRONMENTAL POLITICS Andrew Dobson
EPICUREANISM Catherine Wilson
EPIDEMIOLOGY Rodolfo Saracci
ETHICS Simon Blackburn
ETHNOMUSICOLOGY Timothy Rice
THE ETRUSCANS Christopher Smith
EUGENICS Philippa Levine
THE EUROPEAN UNION Simon Usherwood and John Pinder
EUROPEAN UNION LAW Anthony Arnull
EVOLUTION Brian and Deborah Charlesworth
EXISTENTIALISM Thomas Flynn
EXPLORATION Stewart A. Weaver
THE EYE Michael Land
FAIRY TALE Marina Warner
FAMILY LAW Jonathan Herring
FASCISM Kevin Passmore
FASHION Rebecca Arnold
FEMINISM Margaret Walters
FILM Michael Wood

FILM MUSIC Kathryn Kalinak
THE FIRST WORLD WAR Michael Howard
FOLK MUSIC Mark Slobin
FOOD John Krebs
FORENSIC PSYCHOLOGY David Canter
FORENSIC SCIENCE Jim Fraser
FORESTS Jaboury Ghazoul
FOSSILS Keith Thomson
FOUCAULT Gary Gutting
THE FOUNDING FATHERS R. B. Bernstein
FRACTALS Kenneth Falconer
FREE SPEECH Nigel Warburton
FREE WILL Thomas Pink
FREEMASONRY Andreas Önnerfors
FRENCH LITERATURE John D. Lyons
THE FRENCH REVOLUTION William Doyle
FREUD Anthony Storr
FUNDAMENTALISM Malise Ruthven
FUNGI Nicholas P. Money
THE FUTURE Jennifer M. Gidley
GALAXIES John Gribbin
GALILEO Stillman Drake
GAME THEORY Ken Binmore
GANDHI Bhikhu Parekh
GENES Jonathan Slack
GENIUS Andrew Robinson
GENOMICS John Archibald
GEOGRAPHY John Matthews and David Herbert
GEOLOGY Jan Zalasiewicz
GEOPHYSICS William Lowrie
GEOPOLITICS Klaus Dodds
GERMAN LITERATURE Nicholas Boyle
GERMAN PHILOSOPHY Andrew Bowie
GLOBAL CATASTROPHES Bill McGuire
GLOBAL ECONOMIC HISTORY Robert C. Allen
GLOBALIZATION Manfred Steger
GOD John Bowker
GOETHE Ritchie Robertson
THE GOTHIC Nick Groom
GOVERNANCE Mark Bevir
GRAVITY Timothy Clifton
THE GREAT DEPRESSION AND THE NEW DEAL Eric Rauchway
HABERMAS James Gordon Finlayson
THE HABSBURG EMPIRE Martyn Rady
HAPPINESS Daniel M. Haybron
THE HARLEM RENAISSANCE Cheryl A. Wall
THE HEBREW BIBLE AS LITERATURE Tod Linafelt
HEGEL Peter Singer
HEIDEGGER Michael Inwood
THE HELLENISTIC AGE Peter Thonemann
HEREDITY John Waller
HERMENEUTICS Jens Zimmermann
HERODOTUS Jennifer T. Roberts
HIEROGLYPHS Penelope Wilson
HINDUISM Kim Knott
HISTORY John H. Arnold
THE HISTORY OF ASTRONOMY Michael Hoskin
THE HISTORY OF CHEMISTRY William H. Brock
THE HISTORY OF CINEMA Geoffrey Nowell-Smith
THE HISTORY OF LIFE Michael Benton
THE HISTORY OF MATHEMATICS Jacqueline Stedall
THE HISTORY OF MEDICINE William Bynum
THE HISTORY OF PHYSICS J. L. Heilbron
THE HISTORY OF TIME Leofranc Holford-Strevens
HIV AND AIDS Alan Whiteside
HOBBES Richard Tuck
HOLLYWOOD Peter Decherney
THE HOLY ROMAN EMPIRE Joachim Whaley
HOME Michael Allen Fox
HORMONES Martin Luck
HUMAN ANATOMY Leslie Klenerman
HUMAN EVOLUTION Bernard Wood
HUMAN RIGHTS Andrew Clapham
HUMANISM Stephen Law
HUME A. J. Ayer
HUMOUR Noël Carroll
THE ICE AGE Jamie Woodward
IDEOLOGY Michael Freeden
THE IMMUNE SYSTEM Paul Klenerman
INDIAN CINEMA Ashish Rajadhyaksha

INDIAN PHILOSOPHY Sue Hamilton
THE INDUSTRIAL REVOLUTION Robert C. Allen
INFECTIOUS DISEASE Marta L. Wayne and Benjamin M. Bolker
INFINITY Ian Stewart
INFORMATION Luciano Floridi
INNOVATION Mark Dodgson and David Gann
INTELLIGENCE Ian J. Deary
INTELLECTUAL PROPERTY Siva Vaidhyanathan
INTERNATIONAL LAW Vaughan Lowe
INTERNATIONAL MIGRATION Khalid Koser
INTERNATIONAL RELATIONS Paul Wilkinson
INTERNATIONAL SECURITY Christopher S. Browning
IRAN Ali M. Ansari
ISLAM Malise Ruthven
ISLAMIC HISTORY Adam Silverstein
ISOTOPES Rob Ellam
ITALIAN LITERATURE Peter Hainsworth and David Robey
JESUS Richard Bauckham
JEWISH HISTORY David N. Myers
JOURNALISM Ian Hargreaves
JUDAISM Norman Solomon
JUNG Anthony Stevens
KABBALAH Joseph Dan
KAFKA Ritchie Robertson
KANT Roger Scruton
KEYNES Robert Skidelsky
KIERKEGAARD Patrick Gardiner
KNOWLEDGE Jennifer Nagel
THE KORAN Michael Cook
LAKES Warwick F. Vincent
LANDSCAPE ARCHITECTURE Ian H. Thompson
LANDSCAPES AND GEOMORPHOLOGY Andrew Goudie and Heather Viles
LANGUAGES Stephen R. Anderson
LATE ANTIQUITY Gillian Clark
LAW Raymond Wacks
THE LAWS OF THERMODYNAMICS Peter Atkins
LEADERSHIP Keith Grint
LEARNING Mark Haselgrove
LEIBNIZ Maria Rosa Antognazza
LIBERALISM Michael Freeden
LIGHT Ian Walmsley
LINCOLN Allen C. Guelzo
LINGUISTICS Peter Matthews
LITERARY THEORY Jonathan Culler
LOCKE John Dunn
LOGIC Graham Priest
LOVE Ronald de Sousa
MACHIAVELLI Quentin Skinner
MADNESS Andrew Scull
MAGIC Owen Davies
MAGNA CARTA Nicholas Vincent
MAGNETISM Stephen Blundell
MALTHUS Donald Winch
MAMMALS T. S. Kemp
MANAGEMENT John Hendry
MAO Delia Davin
MARINE BIOLOGY Philip V. Mladenov
THE MARQUIS DE SADE John Phillips
MARTIN LUTHER Scott H. Hendrix
MARTYRDOM Jolyon Mitchell
MARX Peter Singer
MATERIALS Christopher Hall
MATHEMATICS Timothy Gowers
THE MEANING OF LIFE Terry Eagleton
MEASUREMENT David Hand
MEDICAL ETHICS Tony Hope
MEDICAL LAW Charles Foster
MEDIEVAL BRITAIN John Gillingham and Ralph A. Griffiths
MEDIEVAL LITERATURE Elaine Treharne
MEDIEVAL PHILOSOPHY John Marenbon
MEMORY Jonathan K. Foster
METAPHYSICS Stephen Mumford
THE MEXICAN REVOLUTION Alan Knight
MICHAEL FARADAY Frank A. J. L. James
MICROBIOLOGY Nicholas P. Money
MICROECONOMICS Avinash Dixit
MICROSCOPY Terence Allen
THE MIDDLE AGES Miri Rubin
MILITARY JUSTICE Eugene R. Fidell
MILITARY STRATEGY Antulio J. Echevarria II
MINERALS David Vaughan
MIRACLES Yujin Nagasawa

MODERN ART David Cottington
MODERN CHINA Rana Mitter
MODERN DRAMA
Kirsten E. Shepherd-Barr
MODERN FRANCE
Vanessa R. Schwartz
MODERN INDIA Craig Jeffrey
MODERN IRELAND Senia Pašeta
MODERN ITALY Anna Cento Bull
MODERN JAPAN
Christopher Goto-Jones
MODERN LATIN AMERICAN LITERATURE
Roberto González Echevarría
MODERN WAR Richard English
MODERNISM Christopher Butler
MOLECULAR BIOLOGY Aysha Divan and Janice A. Royds
MOLECULES Philip Ball
MONASTICISM Stephen J. Davis
THE MONGOLS Morris Rossabi
MOONS David A. Rothery
MORMONISM
Richard Lyman Bushman
MOUNTAINS Martin F. Price
MUHAMMAD Jonathan A. C. Brown
MULTICULTURALISM Ali Rattansi
MULTILINGUALISM John C. Maher
MUSIC Nicholas Cook
MYTH Robert A. Segal
THE NAPOLEONIC WARS
Mike Rapport
NATIONALISM Steven Grosby
NATIVE AMERICAN LITERATURE
Sean Teuton
NAVIGATION Jim Bennett
NELSON MANDELA Elleke Boehmer
NEOLIBERALISM Manfred Steger and Ravi Roy
NETWORKS Guido Caldarelli and Michele Catanzaro
THE NEW TESTAMENT
Luke Timothy Johnson
THE NEW TESTAMENT AS LITERATURE Kyle Keefer
NEWTON Robert Iliffe
NIETZSCHE Michael Tanner
NINETEENTH-CENTURY BRITAIN Christopher Harvie and H. C. G. Matthew
THE NORMAN CONQUEST
George Garnett
NORTH AMERICAN INDIANS
Theda Perdue and Michael D. Green
NORTHERN IRELAND
Marc Mulholland
NOTHING Frank Close
NUCLEAR PHYSICS Frank Close
NUCLEAR POWER Maxwell Irvine
NUCLEAR WEAPONS
Joseph M. Siracusa
NUMBERS Peter M. Higgins
NUTRITION David A. Bender
OBJECTIVITY Stephen Gaukroger
OCEANS Dorrik Stow
THE OLD TESTAMENT
Michael D. Coogan
THE ORCHESTRA D. Kern Holoman
ORGANIC CHEMISTRY
Graham Patrick
ORGANIZED CRIME
Georgios A. Antonopoulos and Georgios Papanicolaou
ORGANIZATIONS Mary Jo Hatch
PAGANISM Owen Davies
PAIN Rob Boddice
THE PALESTINIAN-ISRAELI CONFLICT Martin Bunton
PANDEMICS Christian W. McMillen
PARTICLE PHYSICS Frank Close
PAUL E. P. Sanders
PEACE Oliver P. Richmond
PENTECOSTALISM William K. Kay
PERCEPTION Brian Rogers
THE PERIODIC TABLE Eric R. Scerri
PHILOSOPHY Edward Craig
PHILOSOPHY IN THE ISLAMIC WORLD Peter Adamson
PHILOSOPHY OF LAW
Raymond Wacks
PHILOSOPHY OF SCIENCE
Samir Okasha
PHILOSOPHY OF RELIGION
Tim Bayne
PHOTOGRAPHY Steve Edwards
PHYSICAL CHEMISTRY Peter Atkins
PILGRIMAGE Ian Reader
PLAGUE Paul Slack
PLANETS David A. Rothery
PLANTS Timothy Walker

PLATE TECTONICS Peter Molnar
PLATO Julia Annas
POLITICAL PHILOSOPHY David Miller
POLITICS Kenneth Minogue
POPULISM Cas Mudde and Cristóbal Rovira Kaltwasser
POSTCOLONIALISM Robert Young
POSTMODERNISM Christopher Butler
POSTSTRUCTURALISM Catherine Belsey
POVERTY Philip N. Jefferson
PREHISTORY Chris Gosden
PRESOCRATIC PHILOSOPHY Catherine Osborne
PRIVACY Raymond Wacks
PROBABILITY John Haigh
PROGRESSIVISM Walter Nugent
PROJECTS Andrew Davies
PROTESTANTISM Mark A. Noll
PSYCHIATRY Tom Burns
PSYCHOANALYSIS Daniel Pick
PSYCHOLOGY Gillian Butler and Freda McManus
PSYCHOTHERAPY Tom Burns and Eva Burns-Lundgren
PUBLIC ADMINISTRATION Stella Z. Theodoulou and Ravi K. Roy
PUBLIC HEALTH Virginia Berridge
PURITANISM Francis J. Bremer
THE QUAKERS Pink Dandelion
QUANTUM THEORY John Polkinghorne
RACISM Ali Rattansi
RADIOACTIVITY Claudio Tuniz
RASTAFARI Ennis B. Edmonds
THE REAGAN REVOLUTION Gil Troy
REALITY Jan Westerhoff
THE REFORMATION Peter Marshall
RELATIVITY Russell Stannard
RELIGION IN AMERICA Timothy Beal
THE RENAISSANCE Jerry Brotton
RENAISSANCE ART Geraldine A. Johnson
REVOLUTIONS Jack A. Goldstone
RHETORIC Richard Toye
RISK Baruch Fischhoff and John Kadvany
RITUAL Barry Stephenson
RIVERS Nick Middleton
ROBOTICS Alan Winfield
ROCKS Jan Zalasiewicz
ROMAN BRITAIN Peter Salway
THE ROMAN EMPIRE Christopher Kelly
THE ROMAN REPUBLIC David M. Gwynn
ROMANTICISM Michael Ferber
ROUSSEAU Robert Wokler
RUSSELL A. C. Grayling
RUSSIAN HISTORY Geoffrey Hosking
RUSSIAN LITERATURE Catriona Kelly
THE RUSSIAN REVOLUTION S. A. Smith
SAVANNAS Peter A. Furley
SCHIZOPHRENIA Chris Frith and Eve Johnstone
SCHOPENHAUER Christopher Janaway
SCIENCE AND RELIGION Thomas Dixon
SCIENCE FICTION David Seed
THE SCIENTIFIC REVOLUTION Lawrence M. Principe
SCOTLAND Rab Houston
SEXUAL SELECTION Marlene Zuk and Leigh W. Simmons
SEXUALITY Véronique Mottier
SHAKESPEARE'S COMEDIES Bart van Es
SHAKESPEARE'S SONNETS AND POEMS Jonathan F. S. Post
SHAKESPEARE'S TRAGEDIES Stanley Wells
SIKHISM Eleanor Nesbitt
THE SILK ROAD James A. Millward
SLANG Jonathon Green
SLEEP Steven W. Lockley and Russell G. Foster
SOCIAL AND CULTURAL ANTHROPOLOGY John Monaghan and Peter Just
SOCIAL PSYCHOLOGY Richard J. Crisp
SOCIAL WORK Sally Holland and Jonathan Scourfield
SOCIALISM Michael Newman
SOCIOLINGUISTICS John Edwards
SOCIOLOGY Steve Bruce
SOCRATES C. C. W. Taylor
SOUND Mike Goldsmith
THE SOVIET UNION Stephen Lovell
THE SPANISH CIVIL WAR Helen Graham
SPANISH LITERATURE Jo Labanyi
SPINOZA Roger Scruton

SPIRITUALITY Philip Sheldrake
SPORT Mike Cronin
STARS Andrew King
STATISTICS David J. Hand
STEM CELLS Jonathan Slack
STOICISM Brad Inwood
STRUCTURAL ENGINEERING David Blockley
STUART BRITAIN John Morrill
SUPERCONDUCTIVITY Stephen Blundell
SYMMETRY Ian Stewart
SYNTHETIC BIOLOGY Jamie A. Davies
TAXATION Stephen Smith
TEETH Peter S. Ungar
TELESCOPES Geoff Cottrell
TERRORISM Charles Townshend
THEATRE Marvin Carlson
THEOLOGY David F. Ford
THINKING AND REASONING Jonathan St B. T. Evans
THOMAS AQUINAS Fergus Kerr
THOUGHT Tim Bayne
TIBETAN BUDDHISM Matthew T. Kapstein
TOCQUEVILLE Harvey C. Mansfield
TRAGEDY Adrian Poole
TRANSLATION Matthew Reynolds
THE TROJAN WAR Eric H. Cline
TRUST Katherine Hawley
THE TUDORS John Guy
TWENTIETH-CENTURY BRITAIN Kenneth O. Morgan
THE UNITED NATIONS Jussi M. Hanhimäki
UNIVERSITIES AND COLLEGES David Palfreyman and Paul Temple
THE U.S. CONGRESS Donald A. Ritchie
THE U.S. CONSTITUTION David J. Bodenhamer
THE U.S. SUPREME COURT Linda Greenhouse
UTILITARIANISM Katarzyna de Lazari-Radek and Peter Singer
UTOPIANISM Lyman Tower Sargent
VETERINARY SCIENCE James Yeates
THE VIKINGS Julian Richards
VIRUSES Dorothy H. Crawford
VOLTAIRE Nicholas Cronk
WAR AND TECHNOLOGY Alex Roland
WATER John Finney
WEATHER Storm Dunlop
THE WELFARE STATE David Garland
WILLIAM SHAKESPEARE Stanley Wells
WITCHCRAFT Malcolm Gaskill
WITTGENSTEIN A. C. Grayling
WORK Stephen Fineman
WORLD MUSIC Philip Bohlman
THE WORLD TRADE ORGANIZATION Amrita Narlikar
WORLD WAR II Gerhard L. Weinberg
WRITING AND SCRIPT Andrew Robinson
ZIONISM Michael Stanislawski

Available soon:

ARTIFICIAL INTELLIGENCE Margaret A. Boden
AMERICAN CULTURAL HISTORY Eric Avila
THE BOOK OF COMMON PRAYER Brian Cummings
MODERN ARCHITECTURE Adam Sharr
SEXUAL SELECTION Marlene Zuk and Leigh W. Simmons

For more information visit our website

www.oup.com/vsi/

Jan Zalasiewicz

GEOLOGY

A Very Short Introduction

OXFORD
UNIVERSITY PRESS

Great Clarendon Street, Oxford, OX2 6DP,
United Kingdom

Oxford University Press is a department of the University of Oxford.
It furthers the University's objective of excellence in research, scholarship,
and education by publishing worldwide. Oxford is a registered trade mark of
Oxford University Press in the UK and in certain other countries

First edition published in 2018

Impression: 1

Published in the United States of America by Oxford University Press
198 Madison Avenue, New York, NY 10016, United States of America

British Library Cataloguing in Publication Data
Data available

Library of Congress Control Number: 2018936922

ISBN 978–0–19–880445–1

Printed in Great Britain by
Ashford Colour Press Ltd, Gosport, Hampshire

To my colleagues in geology

Contents

Preface

This book is a sketch of an enormous, multifaceted subject, seen through the lens of my own lifelong journey through the subject, and of the enthusiasms that happened to develop along the way (and the nice thing about the subject is that few of its aspects did not become enthusiasms, once encountered). I hope that it provides some kind of introduction to the spirit and the variety of geology.

I'd like to thank Latha Menon, Jenny Nugee, Sandy Garel, Joy Mellor, Dorothy McCarthy, Elakkia Bharathi, and their colleagues at Oxford University Press, together with the readers of the initial manuscript, who skilfuly and patiently helped steer the narrative, as words and pictures, into its final state. For the provision of images, my grateful thanks to Grażyna Kryza (for photos taken by the late Ryszard Kryza), Mark Williams, Annika Burns, Latha Menon, and the British Geological Survey. I'd like to thank too, all the friends and colleagues who have shared various stages of my life with geology, from initial childhood inspiration, where the curators of Bolton and Ludlow museums played a rather larger role than they may have imagined, through to my mentors at Sheffield and Cambridge universities where the studies became serious, and then my colleagues first at the British Geological Survey in the early part of my career and now at the University of Leicester. They have all been crucial in shaping my understanding of geology.

List of illustrations

Chapter 1
What is geology?

To do geology is a little like having the world's biggest and best time machine at one's disposal. The aspect of time is, of course, built in. We have as our field of enquiry—as that notable geologist and palaeontologist Richard Fortey has said—the whole Earth, and its 4.54-billion-year history, and everything that has formed on our planet over that enormous span of time (Figure 1). The early geologists of the 19th century were entranced, and awed, and mystified by the fossilized bones of monstrous creatures, unlike anything alive today, that they unearthed from sea cliffs and quarries. The sense of wonder has not diminished as our collections of dinosaurs and other extinct organisms have grown. Indeed, it has deepened as new discoveries have continued to be made—of flying reptiles with 12-metre wingspans, of feathered dinosaurs, and of whole new families of fossil invertebrates. The level of detail that may survive underground through very many millions of years can be astounding, too—of not just bones and shells, but of skin, eyes, and feathers; of microscopic details of cellular tissues; and even of fragments of DNA.

The ancient worlds inhabited by the dinosaurs and other long-vanished organisms are also part of geology. The swamplands and river floodplains on which the dinosaurs roamed can be as beautifully preserved as are their fossil remains. Those rock strata in which their bones are entombed are, in the most real sense, the

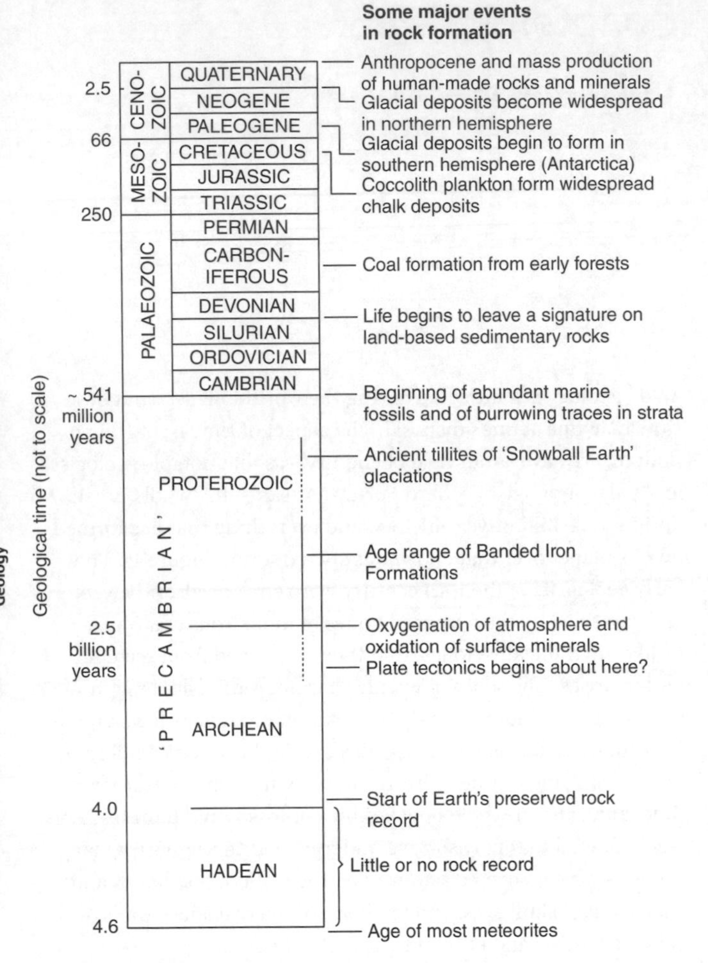

1. Geological timescale.

physical remains of ancient landscapes from which a palpable sense of their environment can be gleaned, including some of the most subtle of events. Imprints of individual raindrops, the skittering of tiny insect legs (as well as the heavy tread of dinosaur feet), microscopic snowfalls of pollen grains—all of these can be imprinted into the rock surfaces. And because these are *strata*, and not just the surface of one rock layer, they represent countless landscapes, building up on top of each other. The real geological record is that of a dynamic and evolving landscape, quite unlike the still-life tableaux of old museum exhibitions.

And for the most part, these are mostly not landscapes, but rather submarine-scapes—the physical remains of the bottom of the sea. They can faithfully preserve the evidence of phenomena that humans have never (even now) more than glimpsed, such as the traces of catastrophic sediment flows that transported many billions of tons of sediment for thousands of kilometres. It is geology, and the rock record, that make these processes—alien to land-living creatures—vividly apparent.

Our time machine, though, does not just take us to long-vanished landscapes and ocean floors. Like one of the more outlandish science fiction machines, it can pass through the ground itself, taking us through into zones where strata are compressed by the enormous weight of sediment above and simmered by the heat of the Earth below, so that oil and gas is wrung out of them. This underground journey can go further, passing through the roots of mountain belts where the rocks, squeezed by enormous tectonic forces and heated to near melting, are now unrecognizable from the fossil-rich strata they used to be. And, where the rocks do melt, we can visit the magma chambers that form, and then stay with them as they slowly solidify—or follow those parts of the magma that race to the surface to erupt, where both the volcano's complex fractured interior as well as its familiar surface topography are within a geologist's gaze. The journey can plunge further down, all the way to the core of the Earth. The glimpses we have of this realm are shadowy and fitful—but they do reveal processes acting on a planetary scale, like the seismic images of tectonic plates slowly sinking into the depths of the Earth's mantle, or of rising plumes of mantle material, thousands of kilometres high.

Our geological time machine can cross dimensions in other ways, into realms where even science fiction writers rarely tread. Today, one can journey into the heart of a single tiny crystal—that grew, say, inside a magma chamber—and examine the way it grew, layer by microscopic layer, in patterns that reflect the physical and chemical 'weather' formed by the magma currents. Some individual, highly resistant crystals, interrogated by carefully directed beams of electrons or ions, can show extraordinary stories of travel—erupted out onto the Earth's surface, where they may spend a few billion years resting in sedimentary layers, before finding their way into another magma chamber to grow a little more, and then being released again, to end their journey under a geologist's microscope. Some tiny crystals tell much larger stories. A few years ago, a tiny fragment of mineral found *within* a diamond was discovered to have formed in the crushing pressures some 500 kilometres down in the Earth's mantle, and survived the journey to the surface within its precious capsule; this mineral fragment revealed that the deep Earth contains, dissolved within it, at least an ocean's-worth of water (Figure 2).

2. A 3-millimetre diamond from Brazil that contains a tiny mineral inclusion betraying the presence of water deep underground in the Earth's mantle.

This kind of science is high adventure in the exploration of the Earth, and that adventure, too, is now taking place on other moons and planets. The explorers today are not humans in space suits with geological hammers and notepads, but sophisticated spacecraft launched into outer space; their cameras and sensors show strange and beautiful geological vistas of startling diversity. Much of the frozen surface of Mars is a landscape from three billion years ago, far older than any Earth landscape. Venus, by contrast, is continually being resurfaced by volcanic eruptions (Figure 3)—some of its lava flows, insulated by the oven-hot atmosphere, are thousands of kilometres long. Even this hellish planet, though, does not take the palm for volcanism. That award must go to Io, the moon of Jupiter, with continual eruptions powered as it is squeezed and kneaded by the tidal forces generated by its massive parent planet. Farther out, Titan, the moon of Saturn, has hydrocarbon rivers and oceans on a water ice landscape—beneath which lies a deep water ocean.

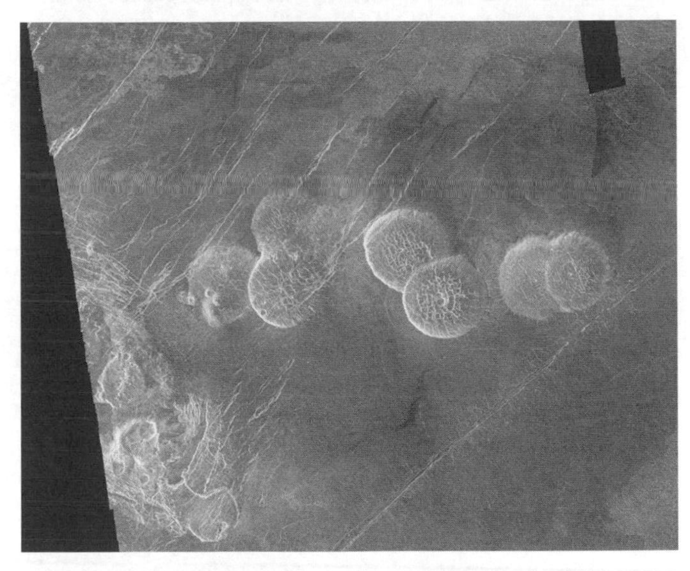

3. The 'pancake volcanoes' of Venus: they are flat-topped and about 25 kilometres across, formed by viscous magma oozing up on to the surface.

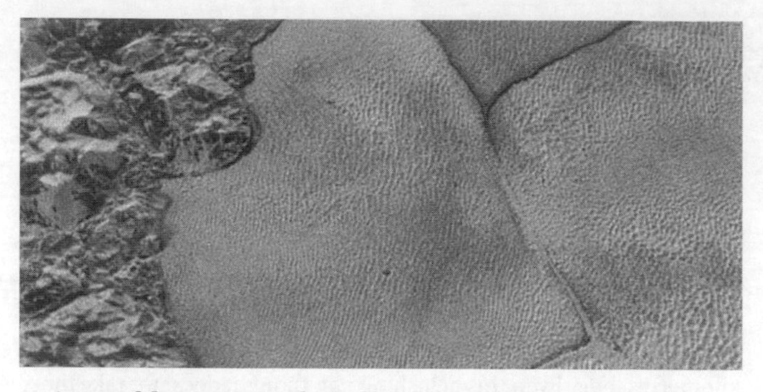

4. A part of the strange geological landscape of Pluto, about 100 kilometres across. Patterned plains of frozen nitrogen abut against craggy hills of water ice.

Even more distant, Pluto, now classed as a dwarf planet, revealed totally unexpected geological formations when the New Horizons spacecraft sped by in 2015: patterned plains of nitrogen ice looking like gigantic quilts lap against rugged ice mountains (Figure 4), while parts of the land surface show a unique 'dragon scale' pattern of high ridges and deep hollows. This diversity represents just the geology of our own solar system. As we begin to glimpse the planets of other star systems, including the rocky 'super-Earths' now known to abound, we can only be sure that many more surprises lie ahead.

The frontiers of geology, hence, extend in all directions, and are nothing if not dramatic. But this science also lies much closer to home. It permeates our lives to the extent that we can be constantly surrounded by geology—yet barely notice it day to day. We live in houses that are made of the reconstituted and flash-metamorphosed mudstone that we call brick, and work in the giant sandcastles held together by lime and mud that we call our concrete buildings. The frameworks of our buildings and of our machines are mostly of iron and steel, almost all of which originated when our planet underwent a protracted chemical revolution between two and three billion years ago. Our machines also include huge amounts of iron, copper, lead,

5. The Bingham Canyon Mine in Utah, USA, dug to extract copper, gold, silver, and molybdenum, is by some measures the largest quarry in the world at 4 kilometres across and nearly 1 kilometre deep.

tin, zinc, aluminium, all of which we extract from ore deposits (Figure 5), often deep underground, while new technology now requires the geologist to prospect for ores of neodymium, gallium, europium, and a host of other less well-known elements, all of which occur in their own particular geological circumstances. Luckily our planet includes a wealth of minerals that is almost certainly unrivalled in our solar system, and most of these minerals are essentially by-products of the (so far as we know) unique planetary engineering that makes up plate tectonics.

Our new urban empires constantly need vast amounts of energy to power them—and this comes from geology, in the shape of the oil and coal and gas that we still utterly rely upon. And as is now all too clear, the burning of these hydrocarbons is changing the chemistry of atmosphere and oceans, and beginning to transform our climate into a new and (for us as for many other organisms) uncomfortable and dangerous state. Again, it is the geological analysis of the climate history of the Earth, of prehistoric episodes of ice ages and of global warming, that allows us to understand how our own species' experiment with this planet compares in scale and speed with past climate perturbations.

Geology, thus, has enormous fascination as a science and huge importance for society. Yet, it is not as well-known as it should be. To many people, the word can still conjure up impressions of rows of dull rock specimens in museum cabinets. Nonetheless, geology provokes great loyalty and affection in those who know it, with many amateur enthusiasts as well as those whose career it is. Professional geologists are usually enthusiasts too, and many simply carry on doing geology for fun when they retire.

Geology is a science that encompasses virtually all the other sciences—chemistry, physics, biology, geography, oceanography, and many others—and involves or impinges upon many of the humanities and arts. There are aspects that geologists adore: the fieldwork, whether in the most exotic or prosaic of places (the polished rock slabs used for decoration in our city centres often show exquisite geological structures); the sense of discovery just around the corner; and the way in which lateral thinking and improvisation lie at the heart of its study. Not least also, there is the camaraderie among amateurs and professionals, and teachers and students, within this science.

This VSI represents a personal overview of a lifelong working geologist (and an enthusiast too, like virtually all my colleagues). It cannot hope to encompass the whole field, but I trust it will give some flavour of what geology is today. One way to begin to tell its story is through the discoveries that have built the science. The men and women who pioneered the science realized, as discovery followed discovery, just how ancient is the planet that we live on, and how dramatic its history.

Chapter 2
Geology: the early days

Our early ancestors likely puzzled, in their freer moments, about what the Earth is made of and how it arose. The early history of this kind of questioning has been completely lost—it probably extended back many thousands of years before the beginning of recorded history. But, among the written records of the early, settled civilizations, such as those of the ancient Greeks, Romans, Indians, and Chinese, fragments of this intellectual adventure can be glimpsed.

Early ideas

Among the earliest recorded speculations on the Earth that we might regard as broadly scientific were those of the ancient Greeks. Their curiosity about the workings of the Earth was just one part of a flowering of art, drama, and philosophy, and extended to many phenomena that we would now regard as geological. There was the shape and size of the whole Earth, for instance. Anaximander of Miletus (611–547 BC) made perhaps the first map of the world, with the Mediterranean as the centre of an Earth that he saw as the circular top of a motionless cylinder. A century later, Pythagoras (570–495 BC) suggested that the Earth was spherical, a claim demonstrated by Aristotle (384–322 BC) through an understanding of how solar eclipses worked.

Eratosthenes (275–195 BC), a Greek who became a librarian at Alexandria in Egypt, later calculated the circumference of that spherical Earth to within 1 per cent of the true value by geometry, using measurements of a shadow cast by the sun at noon at two places, the distance between which was known.

Aristotle—a pupil of Plato who in turn had been a pupil of Socrates—was a key figure, whose influence was to last more than a millennium. While not a scientist in a modern sense, he used logic combined with general observations to generate hypotheses, explaining phenomena in terms of natural rather than supernatural causes, though with mixed results when viewed in the light of modern knowledge. His studies ranged widely, but included questions of what we now call geology. Earthquakes, he suggested, were due to the violent release of trapped winds within the Earth, and fossils were the remains of once-living (though rock-encased, he thought) creatures. He saw that rivers could dry up, and so surmised that seas could also disappear and change place with land. Aristotle's prestige was such that his views—which included the Earth being at the centre of the Universe—overshadowed those of others, such as Anaximander and Democritus (460–370 BC), who saw the Earth as just one of many worlds. Aristotle's ideas continued to hold sway for many centuries, and were even adopted as part of the philosophy of the Catholic Church. Arguably, this for long held back further scientific progress, which was ironic in view of Aristotle's championing of empirical deduction as a foundation to knowledge.

Other cultures independently developed ideas of the Earth and its processes. In India of the Vedic Period (*c.*1300–300 BC), texts such as the Puranas included ideas of cycles of creation and destruction on billion-year timescales. Later scholars such as the mathematician Aryabhata (476–550 AD) realized that the rotation of the Earth made the stars seem to move across the sky; correctly interpreted eclipses as shadows cast by the Moon; and precisely predicted eclipses by calculation. In China, the Song Dynasty

(960–1279 AD) saw major scientific advances, facilitated (and preserved) by the discovery and development of printing. A remarkable polymath, Shen Kuo (1032–95), recognized the nature of fossils, and worked out not only that they were once-living organisms, but also deduced changed positions of land and sea from them—and changes of climate, too. Making observations of fossil bamboo in hilly ground where they did not grow, he suggested that formerly there had been wetter conditions in that region.

Geology in enlightenment times of the Western world

While such advances were being made in the East, the Western world was in the intellectual slowdown of the Dark and Middle Ages. As it was emerging from this, Leonardo da Vinci (1452–1519)—artist, engineer, inventor, musician, scientist—considered the workings of the Earth. Like Shen Kuo before him, he recognized the true nature of fossils, as the petrified remains of very ancient organisms: he saw that marine sea-shells were present on mountainsides above the sea. He saw that these fossils were encased *within* the rock strata (at more than one level), rather than just scattered on the surface, and that some were in life position, with the valves joined together. He deduced that they had been laid down on an ancient sea floor, which silted up and was raised high above sea level. These were very 'modern' findings, though Leonardo used them to put together a view of the Earth in medieval terms, as a gigantic, dynamic macrocosm of the human body, with a circulatory system in the form of rivers. The work was less influential at the time than it might have been, because of the 'mirror-writing' code Leonardo used in his notebooks.

A little after Leonardo's time, the word 'geology' was first introduced, by the Italian Ulisse Aldrovandi (1522–1605), though the term (derived from the Greek words for 'Earth' and 'speech') was not to go into general circulation until the 19th century.

Aldrovandi, who became Professor of Natural Sciences at Bologna University, was regarded as the 'father of natural history' by Georges-Louis Leclerc, Comte de Buffon (1707–88), a prominent figure of the European Enlightenment and one of the most brilliant 'savants' of 18th-century France. Indefatigably hard-working and an eloquent writer, Buffon made his reputation through a thirty-six-volume *Natural History* in which he mainly developed early ideas of biological science. Late in life, though, he wrote a short book, *The Epochs of Nature*, which has a claim to being the first scientific, evidence-based geological history of the Earth. His published estimate of its age (based on the cooling rate of heated metal spheres) was 75,000 years, though privately he thought the Earth was millions of years old. In this history, he deduced a 'single-cycle' Earth, which changed its geography, climate, and biology as it cooled from an original molten ball that, he suggested, had been torn from the Sun by a passing comet. Although this, and much else, was disproved by later work, both the scale and detail of his planetary reconstruction were influential, while some of his deductions were inspired, such as working out that coal seams were the remains of buried, compressed tropical swamps.

Working more or less at the same time as Buffon, James Hutton (1726–97) was a Scottish farmer and scientist, who recognized and argued for the existence of 'deep time' (i.e. geological time)—and demonstrated the great age of the Earth. In contrast to Buffon's finite, cooling-based estimate, he thought this to be an endless span, with 'no vestige of a beginning and no prospect of an end'.

Hutton based his ideas on evidence such as *angular unconformities*, where one series of strata cuts sharply across another (Figure 6). He recognized that this phenomenon must demand enormous amounts of time for the deposition and burial of sedimentary layers; then their deformation, uplift, and erosion; and then subsidence and renewed deposition—with the cycle repeated again to result in modern erosion. Hutton needed an energy

6. A geological unconformity at Siccar Point, Berwickshire, Scotland, where James Hutton in the late 18th century recognized the immensity of deep time: vertical ancient strata (foreground and right) represent the eroded roots of a mountain belt, upon which were laid down much younger sedimentary layers (centre-left of photo). It is now known that these two groups of strata differ in age by some 200 million years.

source to drive this formidable Earth machine, and found it in subterranean heat, recognizing the magmatic and intrusive nature of rocks such as granites (that ascended and dragged adjacent rock strata with them). He was thus a 'plutonist' as opposed to the 'neptunists' of that time, such as the German scientist Abraham Gottlob Werner (1749–1817), who thought that granites had crystallized from seawater. Assembling the evidence, Hutton envisaged the *rock cycle*, with strata being produced by erosion of the land, being buried (and in part metamorphosed or melted) and then uplifted again for the next cycle of erosion to take place.

Further insights into the prehistoric Earth came with Baron Georges Cuvier (1769–1832). During the French Revolution, Georges Cuvier (his background was working class rather than aristocratic) was a tutor in rural France. Fascinated by natural history from an early age, he became, after the Revolution, a member of the National Museum of Natural History in Paris, and went on to become the world's foremost anatomist. Cuvier is perhaps best known for firmly establishing the fact of animal extinction (which Buffon had postulated), by showing clear anatomical differences between the extinct mammoth and the modern elephant. This is not as easy as it now sounds. People were aware of fossils such as trilobites and ammonites preserved within rock strata, but wondered whether such organisms might not be still living in the depths of the oceans, a place then essentially unexplored and unknown (and that occasionally yield 'living fossils' such as the pearly nautilus). The world was by then sufficiently well-explored to show, with reasonable confidence, that a large land animal such as the mammoth could not still be alive anywhere on Earth.

Cuvier is also associated with the idea today termed *catastrophism* (though he used the term 'revolutions'). As the fossils of any region were different from the animals and plants that live there today, he proposed that the organisms that they represented had perished in a succession of 'revolutions' affecting entire regions—but not, he thought, the whole Earth—to be replaced by animals and plants migrating in from unaffected regions. Cuvier did not believe in the idea of biological evolution then being developed, even before Darwin, by scientists such as Lamarck. Although he could show how species had become extinct, he was at a loss to say how they originated.

Buffon, Hutton, and Cuvier were all, in their working lives, established and wealthy enough to follow their scientific interests. Some of the foundations of geology were built, though, by working people, who made their discoveries amid difficult personal circumstances. Mary Anning (1799–1847) was born into an impoverished carpenter's family in Dorset. She became, through determination, skill, and courage, a key figure in the

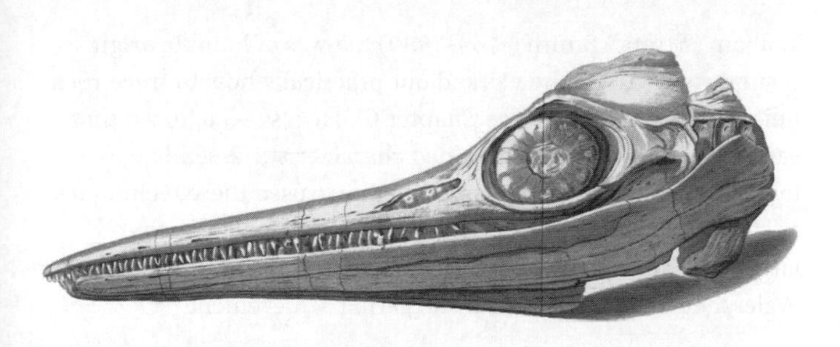

7. The ichthyosaur skull found by Joseph and Mary Anning in 1811 at Lyme Regis, Dorset, and described by Everard Home in 1814 in the *Philosophical Transactions of the Royal Society of London*. It was the discovery and study of fossils such as these that showed that the prehistoric world was inhabited by very different creatures to those living today.

recognition that strange, long-vanished animals had lived in prehistoric time. She excavated, reconstructed, and interpreted some of the first marine reptiles (ichthyosaurs and plesiosaurs) (Figure 7) and flying reptiles (pterosaurs) to emerge from the crumbling and dangerous Jurassic mudstones in the cliffs of Lyme Regis. She had little formal education, yet she formed part of the scientific network of the day, learning French to read Cuvier's publications, and corresponding with leading figures in early geology such as the Reverend William Buckland. Buckland was Professor of Geology at Oxford, later Dean of Westminster, a larger-than-life figure who, among many other accomplishments, named the first-known dinosaur, *Megalosaurus*, in 1824 (and hence beating the gifted amateur Gideon Mantell, who described the *Iguanodon* in 1825, to that honour). Mary Anning provided much of the fossil material that went into reconstructions of ancient environments by Buckland, Cuvier, and others. Charles Dickens was to write about her, and she has been considered 'the greatest fossilist the world ever knew'.

William ('Strata') Smith (1769–1839) too was of humble origins. A surveyor by trade, he worked out practically how to trace rock units across wide areas (see Chapter 6). He also recognized that each of these rock units contained characteristic assemblages of fossils, which could help identify them. He used these techniques to single-handedly, over the course of his life, make the first large-scale geological map of virtually the whole of England, Wales, and Scotland. It was a staggering achievement.

Organized geology

Smith's map today hangs side by side, in the main hallway of the Geological Society of London, with a map that borrowed heavily from it (indeed, alas, plagiarized it). It was drawn by George Bellas Greenough, one of the early presidents of the Geological Society. This was the world's first such organization, and a further step in the establishing of geology. Founded in a London tavern on the 13th of October 1807 by the thirteen founding members (of whom Greenough was one), it was soon to become a focal point for the development of the science.

Charles Lyell (1797–1875) was the 498th member of the Geological Society of London, joining in 1819. He went on to become Secretary, Foreign Secretary, and President of that body, although he had decidedly mixed feelings about the last of these, writing with considerable feeling to Charles Darwin: 'Don't accept any official scientific place if you can avoid it... I fought against the calamity of being President as long as I could.' This might have reflected the cares of office, or perhaps his own scientific reflexes vis-à-vis the stated mission of the Society to collect geological facts rather than to theorize. Lyell certainly appreciated facts—but he was also a profound and adventurous thinker, who greatly influenced the course of the science. He influenced—and became a friend of—Darwin, who took Lyell's ground-breaking book *Principles of Geology* as essential reading on his voyage around the world on the *Beagle*.

Lyell provided opposition to Cuvier's then-popular catastrophic ideas. He championed a radical alternative to catastrophism, and also to a related 'diluvial' idea then current that explained geology in terms of Noah's Flood. Lyell's idea was *uniformitarianism*, usually summarized as 'the present is the key to the past'. Given enough time, Lyell argued, slow and steady changes could produce enormous and far-reaching changes to the Earth. Rivers could excavate mighty gorges; slow crustal movements could build mountain ranges; and enormous volcanoes could be built up by many eruptions. Today, both Cuvier and Lyell are seen to be partly 'right': uniformitarianism works, more or less, over long stretches of geological time—but sudden catastrophes have also occurred (the end-Cretaceous meteorite impact that likely killed off the dinosaurs, for instance) to profoundly alter the course of the Earth's history.

Lyell argued that geology could be *practically* useful to people. This realization led, in 1835, to the setting up of the 'Ordnance Geological Survey' (now the British Geological Survey), the first of its kind, devoted to mapping and describing the rock and mineral formations of the UK, with particular emphasis on such assets as coal, building stone, and groundwater. Over succeeding decades, geological surveys and geological societies were to be formed across much of the world, and they were to gather geological evidence—and to debate new ideas of the Earth too.

One of these new ideas concerned a chaotic, boulder-packed surface sedimentary layer seen to overlie more coherently arranged ancient strata across much of Europe and North America. This chaotic layer, then termed 'Diluvium', was commonly viewed as evidence of a huge recent flood, the Deluge of the Bible. But, some people living and working in the Alps saw modern glaciers producing similar chaotic boulder-rich deposits, and suggested that the widespread 'Diluvium' was the product of a recent Ice Age. The eminent Swiss scientist Louis Agassiz (who later became a major figure in North American geology) in 1836 went to the

Alps to check out this outlandish new theory. Seeing the evidence for himself, he became a major proponent of glacial theory, and worked hard (against considerable resistance) to see it accepted. Agassiz travelled with William Buckland around upland areas of Britain to examine the evidence there, and Buckland, who had previously supported the Flood interpretation, became an enthusiastic convert to glacial theory (Figure 8). Others—including Lyell and Darwin—were more guarded, and it was not until late in the 19th century that the idea of former Ice Ages was generally accepted.

Building a geological timescale

Whether one accepted Buffon's finite duration of our planet, or the infinite one of Hutton, it was clear from the early days of geology that the Earth's timescale vastly exceeded the human one. How could one build a practical calendar of planetary time? Buffon's answer was to divide his inferred planetary history into seven epochs, each with its own allotted fraction of the 75,000 years he publicly allowed for the age of the Earth. The first epoch began with Earth's formation, and the seventh saw human activity 'assisting' the powers of Nature, with epochs in between representing a cooling Earth, a submerged Earth, volcanic outbursts, a warmer Earth (with 'elephants' in now-frigid regions), and continents separating to their present position. It was a clear and vivid history, but it was not to develop into the Geological Time Scale that we now use. That came from another direction, where the focus came not so much from time and inferred history, but from the rocks.

In the 18th century, the age-old business of mineral extraction from the Earth's crust was becoming more widespread and sophisticated. In their day-to-day practical struggles to make sense of the complexities of rock bodies, a couple of mining specialists, Giovanni Arduino (1714–95) in Italy and Johann

8. (a) The formerly glaciated Conwy valley in North Wales; some 20,000 years ago, this valley would have resembled the present Grossglockner valley and glacier of Austria (b).

Gottlob Lehmann (1719–67) in Germany, independently saw a general pattern, exemplified by Arduino's rock-based scheme of:

- *Quaternary* (surface sediment layers)
- *Tertiary* (softer sedimentary rocks overlying the hard strata)
- *Secondary* (hardened strata overlying the crystalline rocks)
- *Primary* (old, crystalline rocks)

This provided a basis for a practical field classification of rocks, and Arduino's and Lehmann's schemes were modified by the influential Abraham Gottlob Werner to become a working geological timescale for the late 18th and early 19th centuries. Werner might be best known today for being on the wrong side of the neptunist/plutonist debate, but this is an unfortunate memorial. Although often in poor health—he abandoned fieldwork early—he was admired as an inspirational teacher, employing a dialogue-based style to instil enthusiasm for geology; students flocked to his lectures at the Freiberg Mining Academy. Some were to take things further.

The first of the new rungs was put in by a scientific colossus, the Prussian-born Alexander von Humboldt (1769–1859), who as a young man made a legendary journey across South America (his account of this journey was another of Darwin's scientific inspirations on the *Beagle*), synthesizing an enormous range of meticulous observations into a holistic model of the Earth. Before his South American voyage, Humboldt had been one of those enthusiastic students drawn to Werner's teaching at Freiberg, and in his early career he worked as a mining inspector, and successfully so: he transformed the fortunes of the gold mines of Bavaria's Fichtel Mountains. Among his wide range of activities, including carrying out botanical research with Goethe, he travelled through the Jura Mountains in 1795, and realized the thick limestone successions there lay outside of Werner's chronology. Humboldt coined the term 'Jura-Kalkstein' in 1799, which Leopold von Buch (another of

Werner's celebrated Freiberg pupils) formalized as the Jurassic System in 1839.

Another striking limestone unit was to provide the next rung. Jean Baptiste Julius d'Omalius d'Halloy was, as his name suggested, scion of an ancient noble Belgian family, destined for a life of high administration, which he indeed mostly pursued. But, inspired by the writings of Buffon, he undertook geological research for several years (much to his family's displeasure), attending Cuvier's lectures and carrying out extensive fieldwork in the region around Paris. There, he proposed, in 1822, the Cretaceous on the basis of the distinctive chalk (which in Latin is *creta*) strata that are so prominent there.

The lower rungs of the modern timescale mainly stem from some of the key figures revolving around the Geological Society of London. In 1822, a founder member, William Phillips, combined with William Conybeare, one of those 19th-century clergymen who enthused about geology. They established the Carboniferous System in 1822, based on the coal deposits that were already powering the country's industrial transformation. Then came a double act that started productively and ended in acrimony, but nevertheless established four major time periods between them, comprising the bulk of what came to be grouped together as the Palaeozoic Era. Adam Sedgwick (1785–1873) was another clergyman who somehow was made Professor of Geology at Cambridge University while knowing little of the subject. He learnt quickly, though, and befriended Roderick Murchison (1792–1871), a wealthy ex-soldier who had been persuaded by the eminent scientist Humphry Davy to do something better with his time than riding to hounds and shooting; as a result Murchison too fell under geology's spell.

Sedgwick and Murchison tackled the ancient rocks of Wales and England's south-west, which contained no dinosaurs but smaller (though no less alluring or enigmatic) fossils such as trilobites.

In Wales in 1835 they combined to set up the oldest fossil-bearing rocks as the Cambrian System (after Cambria, the classical term for Wales), from Sedgwick's work. Above those strata, there were rocks with a different range of fossils studied by Murchison, and these became the Silurian (after the Silures, an ancient Welsh tribe).

In south-west England, too, both men recognized a yet younger dynasty of fossil-bearing strata, that they named Devonian, after the county of Devon. This last suggestion was strongly contested by other geologists of the day who argued that these rocks could not really be distinguished from Silurian strata. Some years later Murchison set off on an expedition to Russia in search of evidence to resolve what became known as 'the great Devonian controversy'. There, using his diplomatic skills to charm, and gain the support of, the Russian Tsar and his court, he found fossiliferous strata that clearly showed that the Devonian was indeed real. As a bonus, near the city of Perm by the Urals, Murchison found equivalents of a thick suite of unfossiliferous rocks that, in Britain, lay above the coal-bearing strata of the Carboniferous. The Perm rocks were a revelation: they abounded in fossils, which clearly demonstrated another dynasty of life on Earth, which became known as the Permian.

The Palaeozoic was complete—nearly. Sedgwick and Murchison, in their later years, quarrelled bitterly over the boundary between the Cambrian and Silurian, each seeking to include the lion's share of the strata within 'their' system. It took a former schoolmaster who became Professor of Geology at Birmingham, Charles Lapworth (1842–1920) to resolve the argument in 1879, setting up the Ordovician System between them.

As the arguments raged in Britain, a genial-natured geologist, Friedrich August von Alberti (1795–1878), was quietly tracing salt-bearing strata in the Stuttgart region of Germany. It was an important job, as salt for many centuries had done the job that tin cans do for us today, as food preservative. He recognized a tripartite division of rocks associated with salt, using this pattern to locate new underground salt reserves—and, indeed, in 1834, to set up a new geological period, the Triassic. This, together with the Jurassic and Cretaceous above, made up the Mesozoic Era—a useful, larger scale unit of geological time that, together with the Palaeozoic in its modern sense, was invented in 1841 by John Phillips (to whom William 'Strata' Smith was relative, guardian, and teacher).

Thus, by the end of the 19th century, the old 'Secondary' division of geological time had disappeared and been replaced by an array of new units. It was realized, too, that the 'Primary' crystalline unit was made up of igneous and metamorphic rocks that could form at any point in geological time (and such rocks are still forming now). The Tertiary hung on for a long time, as the Age of Mammals that followed the Mesozoic dinosaurs, Charles Lyell subdividing it based on successions of increasingly 'modern' fossil species into epochs such as the Eocene and Pliocene. The Tertiary is now formally discontinued (being replaced by the Paleogene and Neogene Periods) but still leads a vigorous informal existence. The Quaternary remains part of the Geological Time Scale, as the time of the recent Ice Age.

The Geological Time Scale (now more formally termed the International Chronostratigraphic Chart) is still evolving. Just in 2004, a new period, the Ediacaran, was set up below the Cambrian, at the top of the enormous span informally termed the Precambrian. Like the other units, it is part of a scale of rock *and* time, simultaneously. Thus, the Ediacaran System is a physical unit, of rocks that can be hammered and drilled through, that represents the long-vanished geological time of the Ediacaran Period.

That time, though, was a mystery to 19th-century geologists, who had no idea of its real duration. The ability to measure Earth time was one of the revolutions of the 20th century that was to transform geology.

Chapter 3
Modern breakthroughs and revolutions

The early geologists built a detailed *relative* history of the Earth, which in general respects still holds true today. But, they had essentially no idea of how long that history was in years. Estimates ranged from just a few million years to billions of years (or, as James Hutton surmised, the Earth being eternal). Ingenious attempts had been made to measure numerical ages, such as by estimating how long it would take for strata to accumulate, or for the oceans to become salty. A certain J. Middleton published a paper in the first ever volume of the *Quarterly Journal of the Geological Society of London*, in 1845, in which he used the progressive uptake of fluorine by fossil bone as a measure of time. His analyses were meticulously carried out and he took care to use reference material of known age (the bones of an Egyptian mummy, and also of a Greek cat 'from the time of the Second Peloponnesian War'). With this he calculated the age of a fossil bone from an extinct camel from strata of the Oligocene Epoch to be 24,200 years. The Oligocene is now known to range from twenty-three to thirty-four million years ago, so poor Mr Middleton was out by three orders of magnitude! Inspired guesswork could be a better bet, such as when William Buckland hazarded that the ichthyosaurs and plesiosaurs that Mary Anning was exhuming from the Jurassic cliffs of Lyme Regis lived 'ten thousand times ten thousand years' ago. That makes a hundred million years, which, at a little over half of the real age of those ancient sea monsters as we now know, is of very much the right order.

In the late 19th century, a formidable scientist, the Scottish physicist William Thomson, who became Lord Kelvin (and after whom the Kelvin scale of temperature is named), entered the fray. He took Buffon's notion of a cooling and solidifying Earth, and applied rigorous mathematical analysis to it. Given the current heat state of the Earth—one that could still generate magma in large amounts—he pronounced that the Earth could be no older than some forty million years. This view was placed overtly in opposition to the growing geological evidence of enormous thicknesses of rock strata containing a long succession of different life-forms as fossils. This was an uncomfortable misfit—but nevertheless many geologists were impressed by the seemingly incontrovertible nature of Kelvin's argument, founded as it was on basic physical principles.

The problem was resolved to general (but not to Kelvin's) satisfaction, through the discovery of radioactivity. This serendipitous discovery was made in 1896 by the French scientist Henri Becquerel, when he noticed that uranium salts, in the dark, created an image on a photographic plate. It was soon realized that this new kind of radiation, emitted from radioactive elements in minerals, could supply the 'missing' energy that had prevented the Earth from solidifying completely in a (geologically) short space of time.

It was soon also realized that the radioactive decay of one element into another, at a steady rate unaffected by heat or pressure, could provide a kind of 'atomic clock' that would allow rocks to be dated, and therefore allow the age of the Earth to be measured. In the first few years of the 20th century, the American chemist Bertram Boltwood and English physicist Ernest (later Lord) Rutherford obtained the first of such *radiometric dates* of rock samples. Although bedevilled by incomplete understanding of the way in which radioactive elements decayed, these first dates ranged to more than two billion years—far older than Lord Kelvin's calculations had indicated.

The techniques involved were difficult, and the results controversial. It took much arduous work by a young English geologist, Arthur Holmes (1890–1965) to develop the methodology, and to produce more reliable dates of rocks that could be tied, using fossils, to the Geological Time Scale. In 1927 Holmes published a book, *The Age of the Earth*, with a timescale that reached back three billion years. Dating the origin of the Earth was difficult, though, because so few rocks or minerals have survived from our planet's beginnings. However, the dating of meteorites—debris left over from the formation of the planets—later indicated an age of more than four and a half billion years for the Earth. It was not just the great age of the Earth that became a revolution in people's minds, but how that time was divided up. Most of Earth time turned out to belong to the Precambrian, with all the familiar periods from the Cambrian onwards making up only some 12 per cent of our planet's history.

Radiometric dating is now an everyday technique in geology, exploiting a range of radioactive elements and their decay products (Figure 9). These include elements such as uranium with very long *half-lives* (the time it takes for half of an amount of the radioactive element to decay into its *daughter product*), of millions or even billions of years. There are other radioactive elements with much shorter half-lives: that of the well-known carbon-14 is measured in thousands of years. This *numerical dating*, when combined with the *relative dating* provided by fossils, provides a means to calibrate the long and eventful history of our planet.

The accumulation of detail

Not all advances in the science are ground-breaking shifts in ruling paradigms, such as the concepts of deep time and of

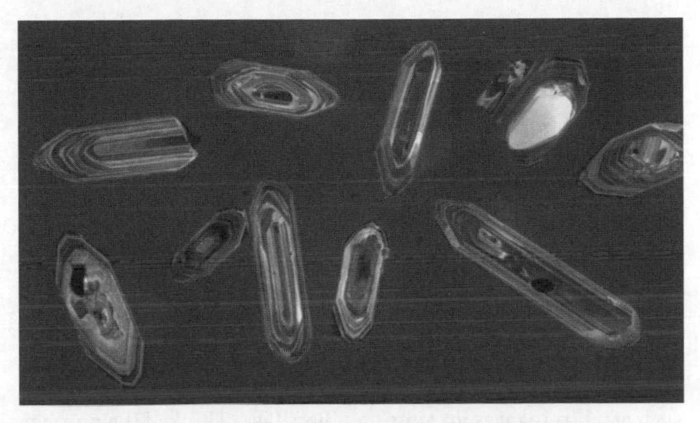

9. Zircon crystals of the kind used for radiometric dating of rocks, due to the uranium they absorbed as they grew, and that has since been decaying into lead.

widespread past glaciations. Much key progress took place in the late 19th and early 20th centuries, as a proliferation of local and regional studies. These took the grand frameworks of geological thought and the Geological Time Scale, and elaborated them, through describing and classifying rock structures and fossil successions in enormously more detail than in the sweeping studies of the early pioneers.

The moving spirit of the Geological Society of London nicely exemplifies the philosophy of those times, which was to catalogue and describe the mass of factual evidence to be found in the rocks, rather than to indulge in fanciful speculation. The ideas and the evidence went hand in hand, of course: nevertheless, the emphasis was on solid, rigorous description.

The protagonists here make up an army of dedicated, tireless, and now largely forgotten individuals, scattered across the globe, who literally hammered a mass of detailed and diverse narratives of Earth process and history from the rocks where they lived and worked. The platform of solid knowledge that they constructed was an indispensable prerequisite to the revolutions in geology that were to follow.

This kind of work followed hard on the heels of the visions of the early pioneers. Buffon, in the late 18th century, in noting petrified remains of animals and plants which seemed no longer to exist on Earth, suggested that it would be a good idea to study these systematically—thus foreshadowing the science of palaeontology. By the early 19th century, the likes of William Smith and Cuvier were already assembling these fossils into a succession of general assemblages, each typifying a different segment of prehistoric time. But the sheer number of different types of fossils meant that such cataloguing was an almost endless task. Think of the number of types of organisms existing today—most still unrecognized and undescribed—and multiply that number many times as one goes back ever further into geological time. Not all species that lived have been fossilized, of course: nevertheless, the scale of description can be daunting.

Among those undaunted was Alcide d'Orbigny (1802–57), son of a ship's doctor, who as a boy living by the coast at La Rochelle developed a precocious interest in marine life there, notably the tiny, shelled foraminifera (amoeba-like protozoa that secrete minute, exquisite shells). He went on to take part in a scientific voyage to South America to collect natural history specimens (preceding Charles Darwin's visit there, who complained that d'Orbigny would have taken all the best specimens!). When he returned, he began systematically studying French fossils, describing nearly 18,000 species and using these to set up fine divisions of strata and time (such as the Toarcian and Oxfordian in the Jurassic, and Albian and Cenomanian in the Cretaceous, subdivisions still in use today).

In Britain, one example was Arthur Rowe, a retired doctor who in the late 19th century studied fossilized sea urchins in chalk strata, recognizing a succession of species that remain a classic example

10. The dramatic landscape of Suilven, as sketched by legendary geologist Ben Peach in the late 19th century, who with John Horne first elucidated the complex rock successions of north-west Scotland. The striking mountain is of sandstones about a billion years old, which rest on much more ancient and altered rocks some three billion years old.

of evolution today (had Darwin known about these fossils that were literally under his feet, he would have been less despairing about fossils as evidence for his theory of natural selection). At about that time, Ben Peach and John Horne were, by painstakingly detailed geological mapping in arduous terrain, unravelling the structures of the ancient mountain belts of north-west Scotland (Figure 10), while in the early 20th century in Wales, Owen Thomas ('OT') Jones was honing similar intuitive skills in the Welsh mountains, producing maps of the rock strata so accurate that they can still guide geologists today.

These were the days, too, of the 'Bone Wars' of the 1870s to 1880s conducted by the palaeontologists Edward Drinker Cope and Charles Othniel Marsh, as they explored, schemed, and fought in bitter competition to find dinosaur bones in the American Mid-West. Amid true skulduggery of the lowest order, they unearthed most of the classic dinosaurs that most vividly express Earth's geological past in the public mind: *Brontosaurus*, *Stegosaurus*, *Triceratops*, *Diplodocus*, and many others. More peaceably, Grove Karl Gilbert (1843–1918) was then exploring the equally arduous terrain of the Rocky Mountains, pioneering studies of how the ancient rock formations related to modern landscapes—and also speculating on geology beyond Earth, in analysing how the Moon's craters were formed.

The depths of the Earth were being explored too—not by any kind of fieldwork (notwithstanding Jules Verne's beautifully imaginative but physically impossible *Journey to the Centre of the Earth*), but by analysing how earthquake waves travel. In 1909, the Croatian scientist Andrija Mohorovičić discovered that there was a distinct boundary separating the Earth's crust from the mantle, still called the Mohorovičić Discontinuity, or 'Moho' for short, in his honour. Three years beforehand, in England, Richard Dixon Oldham had discovered the Earth's core by similar means (Figure 11). As the Earth's anatomy through time and space was being meticulously catalogued, geology was maturing. But, there were still revolutions to come.

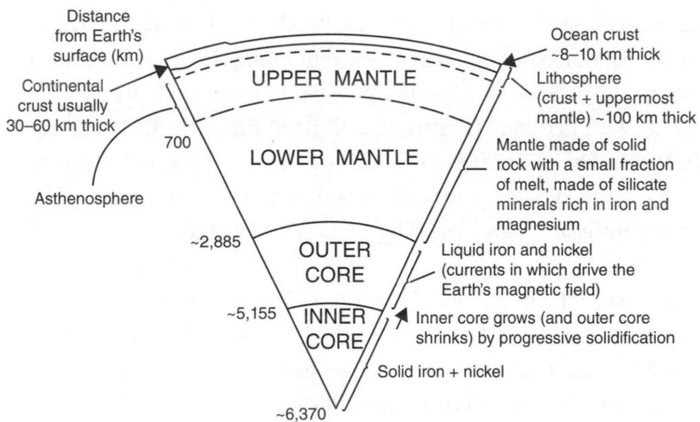

11. The main features of the Earth's core, mantle, and crust.

The puzzle of mountains

On mountain ranges, the Earth's geology is most spectacularly apparent, and so their formation has long been puzzled over. To Buffon, the Earth's mountains were the surface irregularities formed as the Earth cooled from its initial, white-hot state. To Hutton, they were part of an eternal cycle, raised and crumpled by the heat engine of the Earth and then eroded away.

The subsequent detailed study of the Earth's strata soon showed one more puzzling feature. As well as the crumpling of the strata along long, narrow belts—that might be explained, say, by compressive forces at the surface of a cooling and contracting Earth—the very nature of these mountain-strata was distinctive. Rock successions were found to be of modest thickness on the flatter terrain either side of mountain belts. Once inside the mountain belts, though, these strata became not only crumpled and torn apart by tectonic dislocations, but they were also massively increased in thickness. So, prior to the mountain belt ever existing, the nature of the sea floor must also have been quite distinct: to be primed to be a future mountain belt, as it were.

These long narrow areas of ancient sea floor came to be called geosynclines. It was thought that, for many millions of years, geosynclines subsided to a much greater extent than did the sea floor on either side, and so filled with enormous thicknesses of sediment. Then, for reasons that were mysterious but much puzzled over, they stopped subsiding and were forced upwards and compressed to become mountain belts. The geosyncline theory became the standard means of explaining much of the Earth's surface structure. The term was soon elaborated: eugeosynclines were those central parts made up of strata originally formed in deep water, and they were neighboured by miogeosynclines that were made up of shallow-water rocks. As the geologists puzzled, and carried on classifying, terms such as orthogeosyncline, zeugogeosyncline, and parageosyncline arose. This model of Earth structure became sophisticated—but contained also the seeds of its own destruction.

The drifting continents

It was, naturally enough, the mapmakers who first noted the way in which the eastern coastline of the Americas matches that of the western coast of Europe and Africa. One of these, the Flemish cartographer Abraham Ortelius (1527–98: celebrated enough to have his portrait painted by Rubens, and admired as a man who 'served quietly, without accusation, wife or offspring'—as was written on his tombstone), suggested that these continents had once been joined, and then torn apart, 'by earthquakes and floods'.

It took almost four centuries for this idea to seriously re-emerge, most notably driven by the German polar physicist and meteorologist Alfred Wegener. Wegener also noted the coincidence in the shape of the coastlines either side of the Atlantic Ocean, but could also put forward more evidence, thanks to new data assembled by those growing legions of hard-working field geologists. One kind of evidence was certain fossils, which could show close similarity between continents that are now widely separated. Another was the internal structure of continents, where the edges of ancient, long-eroded mountain belts could be matched across a modern ocean (Figure 12). In 1912 he published a paper suggesting that, over geological time, the continents had drifted thousands of kilometres across the Earth.

Wegener acquired a few supporters, but mostly his ideas were rejected—or rather howled down, in some quarters—by the geological community. The identical, far-separated fossils were explained by most geologists via the idea of land bridges—ancient connections between continents that rose up, allowing migration of dinosaurs and other creatures, before subsiding beneath the

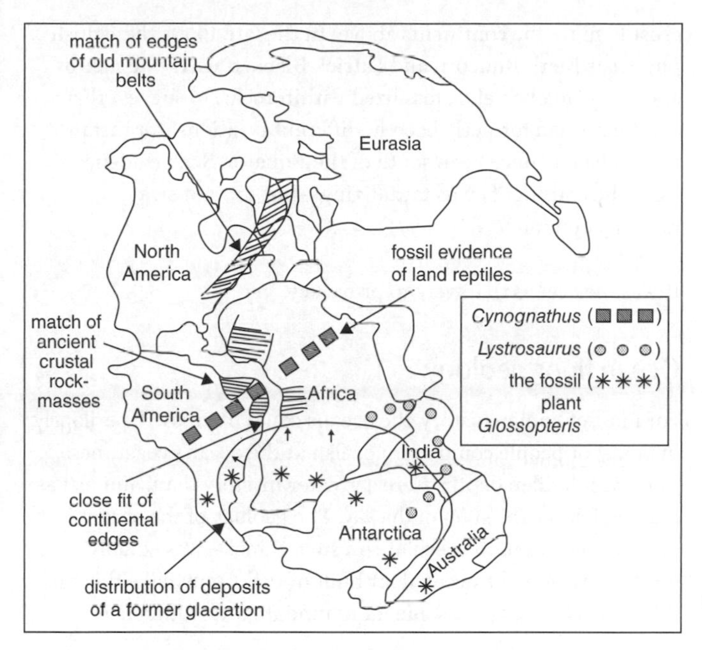

12. Some of the key lines of evidence used by Alfred Wegener to argue for continental drift.

waters again. And the coincidence of the shape and tectonic grain of continents was thought to be just that—coincidence. After all, how could the continents plough through the ocean floor?

Wegener died of a heart attack in 1930, on Greenland's icecap, while his ideas were still heresy. He lies buried there, under some hundred metres of the snow and ice that have fallen since. Of his supporters, Arthur Holmes was the most prominent, and he suggested a motive force for those drifting continents. In his classic textbook, the *Principles of Physical Geology*, which was for the post-Second World War generation what Lyell's *Principles of Geology* was to Victorian-era geologists, he showed a diagram of convection currents on the Earth's mantle dragging at the Earth's crust to move the continents above. In the late 1950s the British physicists Keith Runcorn and Patrick Blackett used evidence of ancient magnetic fields, fossilized within rocks, to suggest that continents had formerly been in different positions: for instance, that India had once been south of the Equator. Such evidence from the continents was tantalizing—but did not sway majority opinion.

It was the oceans that were to prove key.

Ocean floor geology

For most of human history, the oceans could be crossed, perilously, in boats; or people could swim or fish at the ocean's coastlines. Their vast hidden depths were a greater mystery than, and just as inaccessible as, the stars in the sky. The geology of the oceans—once people began to speculate on such things—was equally mysterious. Was the ocean floor built of rocks much like those of the land? Or, was there some more fundamental difference?

Even the depth and shape of the ocean floor was mysterious. It was only in 1839 that the first deep sounding was taken of the ocean floor, when Sir James Clark Ross, captain of the *Terror* and *Erebus*, put together sufficient rope to touch the ocean floor at a depth of 'two thousand four hundred and twenty-five fathoms' (4,435 metres)—a distance beneath sea level as great as the top of Mont Blanc was above it, as that intrepid captain noted. Later in the 19th century, such soundings, laborious as they were (the technology went from rope, to strong twine, to piano wire), began to be used systematically. By the end of the 19th century, the beginnings were glimpsed of a gigantic, ridge-like mountain chain down the middle of the Atlantic.

In the 1920s, sonar technology was developed. This form of echo-sounding was used to hunt enemy submarines, but it could also show how deep the ocean floor was. In the Second World War, one of the US navy ships fitted with this technology happened to be captained by a geologist from Princeton University, Harry Hess, who exploited it to map the shape of the ocean floor where his ship sailed. As well as the enormous submarine mountain ranges, there were also deep trenches that could plunge more than 10 kilometres below sea level. This newly discovered submarine-scape was mysterious. What kind of mountains were these, and how old were they? These mountains were below water, and so beyond the reach of the erosion that wears down and destroys terrestrial mountains. Could they therefore, Hess mused, represent much more ancient topography than anything that was seen on land—far older than the dinosaurs, and perhaps even dating back to the Cambrian Period? Hess was to reject this vision later, but more evidence would be needed to solve the riddle of the ocean floors.

Birth of the plate tectonics concept

More ocean floor evidence was assembled in the 1950s by a pair of innovative geologists, Bruce Heezen and Marie Tharp, in the US, working at the Columbia University's Lamont-Doherty Laboratory. Heezen largely gathered the shipboard data, including systematically made sonar traverses. Tharp, as a woman in those days, was not allowed on board ship; her job was to painstakingly assemble the data by hand, to slowly build up the first detailed map of the ocean floor. As this image began to emerge, she was the first to see something that should not have been there. The familiar mountain ranges on land are typically associated with large-scale compression. But, running down the crest of each of these huge submarine structures there was typically a narrow, steep-sided valley. It looked a lot like a rift valley—a feature that on land was a sign of stretching of the crust, not of compression.

Heezen at first refused to believe it, not least because this unusual combination of features seemed to suggest that the oceans were stretching apart—and that might imply support for the much-denigrated idea of continental drift. He demanded that her maps be re-drawn without the troubling rift valleys. Tharp persisted with her interpretation, and it was to become supported by other lines of evidence. The deep-sea oozes with which the ocean floor was shrouded became thinner, and so seemingly younger, towards the crests of the mid-oceanic mountain ranges. Clusters of earthquake records were shown to be associated with these regions—a sign of tectonic activity. And rock samples suggested that these undersea mountains were made up of the dense iron- and magnesium-rich volcanic rock basalt, unlike the generally less dense silica- and aluminium-rich rocks such as granite that make up terrestrial mountains. The ocean explorer Jacques Cousteau was a sceptic. He set out in one of his deep ocean submersibles, on a mission to disprove Tharp's idea. Once there, astonished, he saw a deep submarine rift valley, exactly where Marie Tharp said it would be.

Heezen was now convinced that the ocean was splitting apart, and creating new crust by the outpouring of basalts along its central line—the mid-ocean ridge. For a while, he thought that must mean that the whole Earth was expanding, puffing up like a balloon. It was Harry Hess, now also convinced of a geologically young rather than ancient ocean, who disposed of this idea. He said that ocean crust was being destroyed at ocean trenches by being pushed down—'subducted'—deep into the Earth's mantle, to balance the crust being created at the mid-ocean ridges, and so to keep the Earth the same size. This is essentially the modern view of *plate tectonics*, in which the continents do not so much plough through the ocean floor, but are carried along with the continually wandering tectonic plates of which the Earth's crust is composed, sometimes splitting apart, sometimes colliding to raise up mountain ranges, and sometimes sliding past each other—as at the San Andreas Fault in California.

Other confirmatory evidence soon turned up, notably magnetic 'stripes' in the ocean crust, symmetrically arranged around the mid-ocean ridges, in which the basalts, as they formed, preserved the pattern of periodic reversals in the Earth's magnetic field, when the North and South poles flipped and changed position. Today, satellite-borne lasers can measure the slow movement of the plates—for instance, the rate at which the Atlantic Ocean is widening, at about 2 centimetres a year, or about as fast as one's fingernails grow.

Plate tectonics made sense of the geosyncline concept. The 'eugeosynclines' turned out to be the ocean trenches, and the thick masses of sediment that accumulated in them before they were squeezed, sheared, and contorted between converging plates; while the 'miogeosynclines' were the shallow-water sedimentary successions bordering them. The new paradigm of plate tectonics shed light on more than just the pattern of mountain ranges. It explained the pattern of volcanism and earthquakes, too. Basalts erupt more or less peacefully along the length of the mid-ocean ridge, while more viscous, silica-rich magmas rise from the region above descending tectonic plates, to breach the surface as the paroxysmal, destructive eruptions typical of regions such as the Pacific 'Ring of Fire'. This is where subduction zones fringe the enormous but now slowly contracting Pacific Ocean, and where the friction-ridden descent of ocean crust is accompanied by powerful earthquakes, which can generate destructive tsunamis (Figure 13).

Plate tectonics is the reason why our planet is one of two halves. The continents, and the shallowly submerged continental shelves, are made of low-density, ancient crust, which is in places well over three billion years old. Despite being slowly eroded by the wind and the rain, the continents are in essence indestructible and will persist for as long as our planet exists. The ocean crust, by contrast, formed of denser basaltic crust, is continually being formed and recycled; only a few small parts of it are more than 200 million years old, and much is less than a hundred million years in age. Despite the youthfulness of the oceans, their detailed

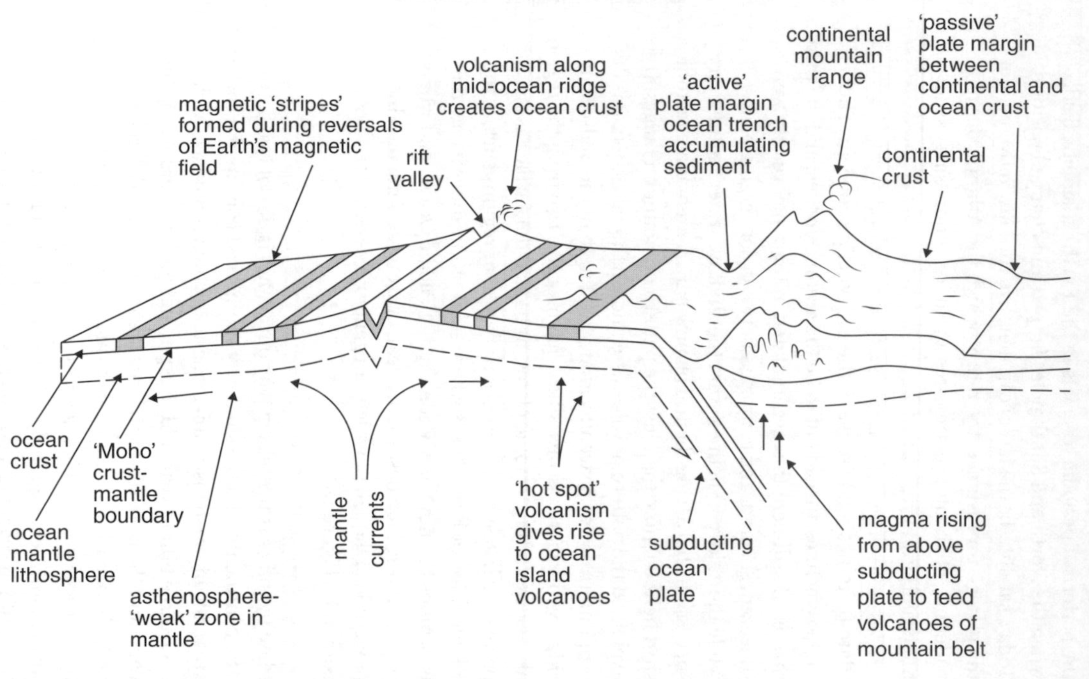

13. **Some of the main features of plate tectonics.**

study is one of the great unsung revolutions in geology. This revolution, though, had a most quirky beginning.

Drilling the oceans

In 1952, a couple of inventive and humorously-minded geophysicists, Gordon Lill and Carl Alexis, were trying to classify a stack of applications at the US Office of Naval Research. Deciding that they could do no better than call them all 'miscellaneous', they went on to form the American Miscellaneous Society, or AMSOC, which became something of a cult institution. Although mainly cherished as an outlet for scientific high spirits, AMSOC was the moving force behind a hugely ambitious project, called the Mohole Project, to drill through the Earth's crust into the mantle. As ocean crust is much thinner than continental crust ('only' being some 10 kilometres thick as opposed to 30–40 kilometres or more) the drilling was to be carried out from a floating platform, with the drill string first needing to penetrate more than 3 kilometres of ocean water. In the early 1960s, this was challenging technically, and the fact that some 180 metres of ocean floor sediments were drilled through, to reach the underlying ocean floor basalts, was a signal achievement. Though the Moho far beneath the ocean floor was not reached (and to this day has not been drilled into), the project was deemed enough of a success to lead to the Deep Sea Drilling Project.

This international project started in 1968, with its own dedicated drilling ship, the *Glomar Challenger*. It and its successor from the 1980s, the *JOIDES Resolution*, have sailed the world's oceans, drilling thousands of boreholes from polar waters to the Equator, often in highly challenging conditions. Along the way it morphed into the Ocean Drilling Program and then into the Integrated Ocean Drilling Program. The task remained the same: to sample that part of our planet's geology that is normally hidden from our view, beneath the deep ocean waters. In its early phases it has confirmed that plate tectonics is real—for instance by showing that the basalts of the ocean crust become older as they become more distant from a mid-ocean ridge.

It went on to provide startlingly detailed histories of the world. The deep ocean oozes are—until their eventual destruction in an oceanic trench, and unlike the patchy, incomplete, and disturbed sedimentary record on land—a continuous and undisturbed archive of the last couple of hundred million years. They have been crucial to working out the history of the Earth's climate, and they have revealed other dramatic episodes in our planet's history, such as the drying out of the Mediterranean Sea six million years ago to leave a 2-kilometre-thick salt layer on its floor. Most recently, boreholes have been drilled directly into the crater left by the meteorite that, sixty-six million years ago, likely killed off the dinosaurs and much else on Earth. Without the adventurous spirit of the American Miscellaneous Society, our knowledge of the Earth might be only half complete.

Chapter 4
Deep Earth geology

We live on the surface of Earth, where air and water meet solid rock, and combine, uniquely in our Solar System, to provide the conditions for life to flourish. But what of the rest of our planet, that lies beneath our feet? The distance from the Earth's surface to the centre of its core is 6,370 kilometres. Humans have penetrated—with difficulty, and with the help of sophisticated technology—to a little more than 4 kilometres below ground, and they have probed with boreholes to a depth of 12 kilometres. That does not even get us through the crust, which proportionately to the whole globe is no thicker than the skin of an apple. So what lies in the gigantic realm of the Earth's depths?

One of the difficulties in our penetrating more deeply into our planet is the heat, combined with the crushing pressure. Buffon in the 18th century had noted the steady increase in temperature on descending deeper into underground mines. The rise in temperature here is something of the order of 3 degrees Celsius every 100 metres. A century later, the writer Louis Figuier, in writing his popular and influential *The World before the Deluge* in 1863, took that rate of temperature increase and simply projected it into the Earth's core: he arrived at a figure of 195,000 degrees! His vision of the Earth as a mass of superhot magma, poised to break through a fragile, pliable crust, was long influential.

Generations of geologists have since puzzled through the evidence of what the subterranean world is really like. It is certainly hot, though the temperature at the Earth's core is now thought to be a more modest 6,000 degrees Celsius, roughly the same temperature as the surface of the Sun. And the magma, which now and then breaks through to the surface in volcanic eruptions, is indeed the basis for one of the main lines of study of the Earth's interior.

Magma and rock

Rocks are the tangible evidence of the Earth's history, and the processes that take place in and on our planet, and therefore their study is a fundamental part of geology. Their study in general is called *petrology*, from the Greek word *pétros* for rock and *lógos* for knowledge (our common use of the word 'petroleum' is to signify oil and gas that are derived from rocks). The part of petrology that deals with rocks that have frozen from their melted state (i.e. from magma) is *igneous petrology* (with *ignis* being the Latin word for fire). Petrology includes two major aspects: *petrography*, which is concerned with describing any rock; and *petrogenesis*, which explores how that rock formed.

Seeing any igneous rock such as a basalt or a granite therefore implies that the rock was once molten, and therefore (in human terms) was at a very high temperature—which for a basalt is something like 1,000 degrees Celsius. A basaltic magma rising from deep underground into cooler regions will begin to solidify, through the growth of crystals of different *minerals*, each type of mineral being a solid substance of fixed chemical composition, or of a composition that varies within fixed limits (and minerals have their own discipline of study, *mineralogy*). Those newly crystallized minerals will not be random, but will depend on the chemical composition of the magma. Common minerals in basalts include *feldspars* (silicon-aluminium oxides with sodium, calcium, or potassium) and *pyroxenes* (silicon-aluminium oxides that can include a range of other chemical elements). If the magma includes enough silica (silicon dioxide, or SiO_2) then the pure form of silica, *quartz*, can crystallize out; if not, then quartz cannot form, but *olivine* (a silicon-aluminium oxide with iron and magnesium) crystallizes instead.

A geologist today can work out the precise mineralogy of such a rock by cutting a very thin, translucent sliver of it (about 30 microns thick) termed a *thin section*, and examining it using a microscope, particularly one fitted with equipment that can polarize the light shining through the thin section (Figure 14). Each mineral has its own optical properties that betray its identity, while the size, arrangement, and relations of the crystals may also be clearly seen. A geologist can also examine this sample under

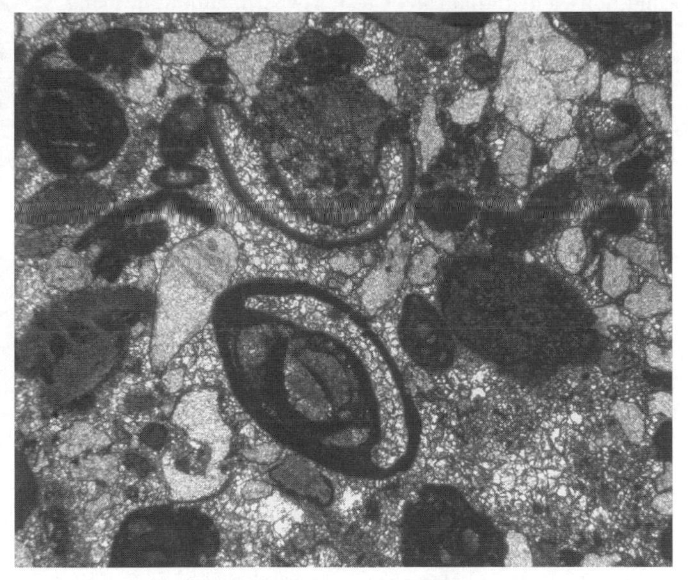

14. A thin section of a 50-million-year-old limestone showing microfossils and mineral and rock fragments, bound together by a finely crystalline natural cement of calcium carbonate.

an electron microscope, or with an *electron microprobe*, in which a finely focused electron beam can 'zap' a tiny area of mineral, as little as two-thousandths of a millimetre across, to determine its chemical composition. Or, the entire rock sample can be crushed and chemically analysed as a whole. What, though, do geologists want to know from this data?

Geologists are often seeking chemical and mineralogical clues to where the original magma came from, and in what conditions of temperature and pressure it formed. The minerals in ocean floor basalts show that the parent magmas formed at depths of just a few tens of kilometres below the sea floor, while diamonds need the pressures associated with depths of at least 140 kilometres to crystallize. Each mineral as it forms has its own *stability field* of temperature, pressure, and chemical conditions, a field that can be tested experimentally, by using machines that can generate enormous pressures and temperatures in a tiny area, to recreate some of the conditions of the deep Earth in the laboratory.

The evidence from rock samples can be put together with very different kinds of evidence, to allow another perspective on the deep Earth.

The travel of waves

When an earthquake strikes as rock masses suddenly shift, the surrounding landscape is shaken sufficiently to crack the ground and topple buildings. The shaking is caused by the passage of different kinds of *seismic waves* through the solid Earth, and these are studied in the discipline of *seismology*. One reason for studying them is to better understand the destructive powers of earthquakes, and to find better ways for society to cope with their effects. Another reason is to use the earthquake waves as a kind of X-ray beam into the deep Earth, to deduce something of the physical properties of the planetary materials that these waves are passing through.

Of the waves spreading out from an earthquake *epicentre*, some travel through the body of the Earth, and others only affect the surface. The body waves come in two forms—the faster-moving primary or 'P' waves, followed by the secondary or 'S' waves. The P waves are pressure waves, much like sound waves travelling through air, and can travel through both solid and liquid material, while the S waves travel by means of a shaking motion, and will only travel through solids. The passage of both of these wave types through all of the Earth's mantle shows that the mantle is essentially solid, and stays solid, despite the high temperatures, because of the effect of high pressure that prevents widespread melting (so where the Earth's crust is cracking apart at the mid-ocean ridges, it is the release of pressure—and not any increase in temperature—that causes melting of the mantle to form the magma that rises to give the basalts of the ocean crust). It is the blocking of the S waves at the base of the mantle that led to the Earth's liquid *outer core* of molten iron and nickel being discovered. And, it was a further subtle variation in the trajectory of P waves deeper still that allowed, in 1938, a solid *inner core* to be discerned, at the centre of the Earth (Figure 15).

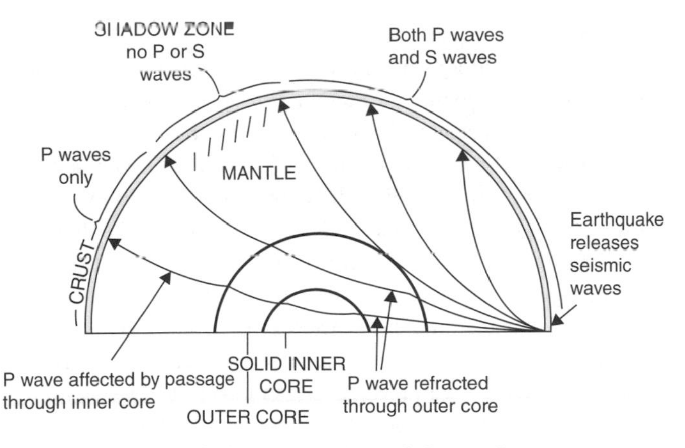

15. The movement of seismic waves through the Earth.

The precise speed and direction of passage of both wave types through the mantle depend on the physical properties of the mantle rocks. Where these are very near melting (or include a small proportion of melted material), the waves tend to slow down. One example of this is a 'weak' layer of rock about 200 kilometres below the Earth's surface. It is this 'weak' layer, called the *asthenosphere*, that acts as a plane of detachment separating the rigid *lithosphere* of the Earth's tectonic plates from the deeper mantle beneath. The tectonic plates, thus, are not made only of the crust but include the uppermost mantle too; this part of the mantle, indeed, makes up most of the bulk of a plate. Waves travelling through the Earth are reflected (rather like the echoes of sound waves bouncing off a wall) and also refracted, changing direction, as they encounter rock bodies of different temperature, pressure, and composition underground.

The complex patterns of seismic waves, as these arrive to different parts of the world, are detected by *seismographs*. Analysis of the wave patterns can be of practical use, in pinpointing the origin or epicentre of an earthquake, and they can also provide much information about the structure of the Earth's interior. Modern seismology uses precise detection of earthquake waves combined with three-dimensional analysis in *seismic tomography*. This relatively new technique has allowed the imaging of some major structures within the mantle. Slabs of lithosphere can be 'seen' descending from ocean trenches, as can the outlines of more or less cylindrical *plumes* of rising, hot (but still essentially solid) rock within the mantle, a little like the shapes formed in old-fashioned lava lamps. These seismological techniques have shown, too, the presence of a thin, complex layer called the 'D-double-prime' (D′′) layer at the core-mantle boundary, which may represent a 'tectonic slab graveyard'. Even subtle details of rock structure can sometimes be discerned, such as whether or not elongated crystals are aligned or randomly arranged within rock masses deep in the Earth. It all adds up to a picture of a deep Earth that is dynamic and complex.

Seismologists looking at the structure of the Earth's crust and uppermost mantle do not always have to wait for an earthquake to strike before being able to gather data. They can create their own shock waves by such means as exploding large amounts of dynamite, and then 'listening' for the echoes from underground.

Matters of gravity

Gravity can also be used to analyse the Earth's structure in the branch of geology generally known as geophysics (of which seismology is part). The effect of gravity, and how it determined the interaction of planets and moons, was famously established by Isaac Newton in 1687, with or without the inspiration of a falling apple (perhaps the story is true, as one of his early biographers, William Stukely, says that Newton told it to him). Newton, however, doubted that the gravitational attraction of a (relatively) small terrestrial structure such as a mountain could be measured.

Newton was proved wrong. In 1738, two French astronomers, Charles Marie de La Condamine and Pierre Bouguer, were on a scientific expedition to South America, to measure the exact length of 1 degree of latitude by the Equator (this was to help determine the precise size and shape of the Earth). They took a lengthy detour to the volcano Chimborazo in Ecuador, and found that they could detect the gravitational attraction it exerted, by the way it very slightly deflected a pendulum. Using the value they obtained they could get an idea of the total density of the Earth. Their results were not precise, but they could at least say that the Earth could not be a hollow shell, as some people then thought.

The expedition had its share of misadventure, and La Condamine and Bouguer later quarrelled and parted company. At the time, La Condamine, a gifted writer and popularizer of science, took most credit for the results. Arguably, though, Bouguer is now more actively remembered, not least every time the gravitational patterns of the Earth are discussed: a 'Bouguer anomaly' is the term given to a slight local variation in the Earth's gravitational field. Such a Bouguer anomaly can be due either to more or less visible mass, like a mountain or a valley, or to invisible features such as high-density (more gravitational pull) or lower-density (less gravitational pull) rock masses that are buried below ground.

La Condamine and Bouguer's experiments were repeated in the late 18th century by more precise measurements of the Scottish mountain Schiehallion, chosen because of its regular, easily measured shape, and because it was made up of more or less uniform rock of known density. Though Schiehallion was not quite as remote as Chimborazo, this measurement too needed arduous work carried out in rough conditions, yet the results were precise enough to allow calculation of the average density of the whole Earth, which turned out to be about double the density of the rocks making up Schiehallion. The Earth's interior must therefore also include masses of very dense rock. We now know this to be the result of the squeezing of silicate minerals into denser forms at depth, together with the nickel-iron composition of the core.

Detailed study of Bouguer anomalies has gone beyond the mapping of static, buried rock masses of different densities. It can now be carried out in real time, by satellite, to show changes in mass due to the melting of icecaps, or to the extraction of water from aquifers. The GRACE (Gravity Recovery And Climate Experiment) programme run by NASA, for instance, very precisely measured the distance between a pair of satellites, orbiting in the same path 220 kilometres apart. As these travelled above areas of larger and smaller gravitational pull on the Earth, they were tugged, exceedingly slightly, backwards and forwards: the resulting minuscule differences in the distance between the satellites were measured and translated into gravity maps of the Earth. Repeated flights by GRACE over the Greenland icecap, for instance, showed that this icecap has lost about 250 billion tons of mass each year since 2002, by melting into the sea.

Magnetic and electric Earth

Two thousand years ago, the Chinese discovered that a magnetic needle aligned itself north–south, and so could be used as a compass for navigation. More than 1,000 years later, the technology spread to Europe. The curious phenomenon of magnetism attracted much speculation, and many thought it was caused by mountains of magnetic material sited on the North pole. The English natural philosopher William Gilbert, though, proposed in a book written in 1600, *De Magnete*, that the whole Earth acted like a bar magnet, with the mass of iron being at the Earth's core—the essence of views that we hold today.

The Earth's magnetic field is not, though, the result of the core being a solid magnetic bar of iron. It was shown in the 1940s by scientists including the English geophysicist Edward Bullard that currents set up in the outer core of liquid iron could act as a dynamo, to generate a magnetic field for the planet. Hence, the North and South poles can slowly 'wander'—change their position slightly—over time (a phenomenon observed by the ancient Chinese).

In geology, magnetism has a number of uses. One exploits the longer and larger scale changes to the Earth's magnetic field, when the North and South poles suddenly flip and change position, that, preserved as the magnetic 'stripes' in the ocean floor crust, were among the primary evidence for plate tectonics. More widely, these 'flips' in the magnetic field are used as time markers in rock successions, as each magnetic reversal was geologically instantaneous around the Earth. As the continents drifted slowly over time, too, their position relative to the North and South poles also changed; that 'apparent polar wander' (because it was the continents wandering, not the poles) was also preserved via the orientation of magnetic minerals acting as fossilized compass needles in the rocks.

Certain rocks contain more magnetic minerals than do others, simply by the nature of their composition (usually those that are more iron-rich). The distribution of more and less magnetic rocks can be detected from the air and displayed as *aeromagnetic maps*. These are a standard and rapid means of assessing the large-scale geological structure of the crust.

The structure of mountains

Extraordinary things happen when mountain ranges are built. Let us take as an example Mount Everest, in the Himalayas, at 8,848 metres the highest peak in the world. Those who make the ascent climb across the frozen remains of rock and magma masses that, over millions of years, made their own journeys within the deep roots of a mountain belt.

The lower parts of the mountain are made of rocks that were pushed deep into the Earth, as the continental mass of India ploughed deep into the heart of Asia. The two continents had first touched fifty million years ago, and India is still moving inexorably forwards today, enlarging the gargantuan dent it has made into Asia. The rocks of what are now Everest's lower slopes were utterly transformed by their passage underground. Most are *metamorphic*—they have been thoroughly re-crystallized, in a hot but still solid state, into rocks such as *gneisses*, the minerals of which, analysed chemically, betray temperatures that then reached 700 degrees Celsius and depths of some 30 kilometres below ground. The fabric of the rock, sheared and crumpled, can be analysed using the techniques of *structural geology* to show that the forces came from north and south, a result of the vice-like grip between the two colliding continents.

Other minerals in these rocks reveal when this took place. At such high temperatures and pressures, crystals of minerals such as zircon (zirconium silicate) and monazite (a phosphate of rare earth elements) can grow. These can take significant amounts of radioactive uranium into their molecular lattice structure, and so can be radiometrically dated. By this means, the Everest metamorphic rocks were shown to have progressively transformed deep underground between about thirty and twenty million years ago. Then something happened.

Hiking around the outskirts of Everest will show that the whole mass of metamorphic rock rests on a gently inclined plane of intensely streaked-out rock. This plane marks where the Everest metamorphic rocks became detached from the underlying rock masses and slid northwards, like a gigantic subterranean river of ductile rock, by some 200 kilometres, in response to the ongoing pressures exerted by the ever-moving Indian continent. The plane is a specific kind of *geological fault*—where rocks break and move past each other due to tectonic forces—termed a *thrust fault*, as rocks are pushed from lower to higher levels along it at a low angle. Some parts of the rock masses were contorted into *geological folds* (Figure 16). About fifteen million years ago, masses of magma, formed in regions where temperatures had risen high enough to melt the rocks, were injected to form masses of granite along planes of weakness. One of these granite masses ballooned spectacularly to a great thickness, and it is this ballooning—a quirk of magma injection—that has hoisted Everest higher than the other surrounding mountains. Since then the forces of erosion have torn away the rock layers above to expose Everest as it now is.

Those who reach the very summit of Everest find one last surprise. Above the mangled metamorphic and igneous rocks, there is another gently inclined tectonic fault plane, above which are limestones, which are scarcely at all metamorphosed. These

16. Tectonic folds of rock strata formed during mountain building. Cap de Creus, Catalonia.

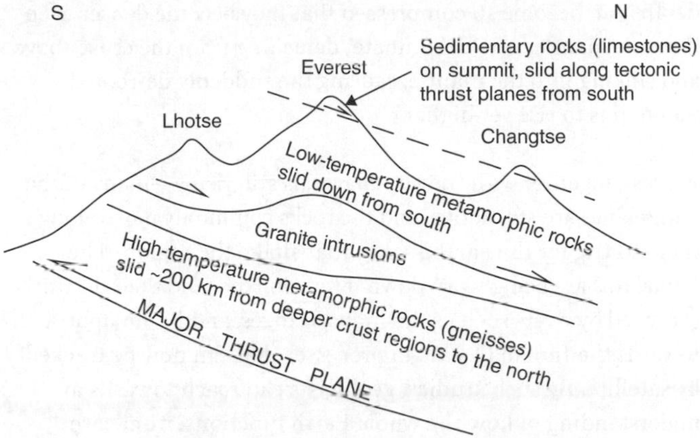

17. The geological structure of Mount Everest.

limestones include fossils of corals and of peculiar tooth-like structures called conodonts, dating from the Ordovician Period, some 450 million years ago. They represent part of an ancient sea floor, from long before the collision, which had never been deeply buried and which slid many kilometres *down* from the south, while the metamorphic rocks below were sliding upwards from the north. It is quite a tectonic sandwich (Figure 17).

These complex rock patterns reflect the larger scale movements of the continental masses, themselves dragged along by the slow-moving currents of mantle rock far below. Geologists can analyse the history of such rocks using clues that range from the microscopic—for instance to see how tiny crystals have been rotated to reconstruct the tectonic forces that acted around them—to the country-scale, as geophysical traverses are made across the mountain range to probe the thickness of the continental crust. In places like the Himalayas, where India is being pushed beneath Asia, the crust can reach 70 kilometres thick, twice as thick as normal continental crust. Such high mountains develop equally impressive roots—which at those depths can become so compressed that they become denser even than the mantle, and 'delaminate', detaching from the crust above and sinking into the mantle, causing the suddenly de-rooted mountains to rise yet higher.

Everest, together with the Himalayas, is still rising—many of the fault planes are still active, and the rocks commonly move along them to trigger the earthquakes that shake the region. The Himalayas are being worn down by erosion too, and their detritus is carried by rivers such as the Indus, Ganges, and Brahmaputra towards the Indian Ocean, in processes that can now be tracked by satellite. By such studies, geologists can reach towards an understanding of how the whole Earth functions, from core to crust. Other kinds of histories reconstruct conditions at the Earth's surface, where it interacts with air and water and life. It is these studies we turn to next.

Chapter 5
Earth surface geology

We are not quite the only body in the solar system with large areas of liquid at the surface. Titan, the largest moon of Saturn, has rivers, lakes, and seas of hydrocarbons such as ethane and methane, amidst a landscape of rock-hard ice that is eroded by nitrogen winds and hydrocarbon rains. NASA's Cassini satellite has imaged wind-blown dunes of grains of ice and solid hydrocarbon and, one murky day in the Xanadu region of Titan (on Titan, all days are murky), on 15 January 2005 of Earth time, the Huygens lander touched down on the pebble-strewn surface of this alien moon.

On present-day Earth, by contrast, there are large areas of ice around the polar regions, but the ice here is soft and warm enough to slowly flow and slide, melting at its edges into the liquid water which fills the oceans and then evaporates into the air to drive an active water cycle, which erodes and decays the silicate rocks at the surface. The process releases nutrients to sustain a biosphere based on hydrocarbons and liquid water, and this living skin is something that—so far as we know—is unique to this planet. It is these particular conditions that, for four billion years or more, have governed the distinctive surface geology of Earth.

Sedimentary world

There are many ways in which the Earth is distinctive, but one of them is certainly the marvellous efficiency with which the *rock cycle* operates. This cycle may be idealized as 'primary' igneous rocks being broken down physically and chemically, by wind, rain, and ice, to produce sediments, which are buried and *lithified* to produce *sedimentary rock*, which becomes metamorphosed as heat and pressure increase, finally ending in melting and magma production—the starting point of another turn of this cycle. Operating on timescales of hundreds of millions of years, this cycle has operated throughout Earth history through the creation and destruction of landscape. This process was intuitively recognized in the late 18th century by James Hutton, and we now see it as driven by the constant operation of plate tectonics.

The segment that is concerned with sedimentary rock is the domain of *sedimentology*. Layers of sediment have been building up on Earth, to form sedimentary rock strata, for at least four billion years. Almost everything we know about the Earth—or at least of the planetary surface that we inhabit—is based on the evidence that we glean from these strata.

The principle is simple—the strata are petrified relics of the land and sea floor surfaces on which they formed. But, in converting them into a prehistoric geography, one's perspective has to be shifted through 90 degrees and then greatly stretched through time. Today, one can walk among modern desert dunes, in the Sahara or Arabian deserts, and see the shapes of the dune surfaces and even—if one has the patience to observe the dunes for a while on a windy day—see them shift their position progressively downwind. But it is hard to see *inside* them: try to dig a hole into them to find out what is there, and the loose sand will soon cave in to leave just a featureless and uninformative hollow.

One can, though, go to examples of fossilized sand wind dunes, for this internal perspective. A classic, beautiful example is the Navajo Sandstone of Utah and neighbouring states in North America, which can be seen in magnificent cliffs, some more than half a kilometre high, in this arid landscape. These dunes are some 180 million years old, dating back to the Jurassic Period when dinosaurs walked the Earth. They have been buried below ground for almost all of that time, where the percolating groundwater has coated the grains in natural cement, binding them together into the hard rock seen in the cliffs. In the rock faces, one only rarely sees dune surfaces as viewed on a modern desert landscape. Rather one sees something more akin to a cinematographic history of the internal structure of a dune as it has migrated through time, appearing as a set of inclined layers (*cross-stratification*) in the rock surface, layers which represent the successive avalanche faces of the moving front of the dune. The tops of these inclined layers are sharply cut off by a near-horizontal erosion surface (a *cross-set boundary*), where the wind blowing down the front of the next train of dunes eroded away the tops of the earlier dunes, before depositing its own inclined sand layers atop their truncated remnants (and then being truncated, eroded, and buried in turn by the next train of dunes) (Figure 18).

The geological record thus shows the process of wind-transport of sand, via fragments of preserved time and remnants of geomorphological features. Nevertheless, from that fragmentary record one can deduce the direction the wind blew from in Jurassic Utah (from the direction of preserved avalanche faces, which point downwind) and even something of the wind speed (from the sizes of the sand grains). One can do this, that is, *if* the Navajo Sandstone really does represent fossilized wind-blown dunes. An alternative idea was once put forward: that the Navajo dunes formed underwater, and were driven by strong tidal currents rather than by wind—as both moving wind and water, despite their very different densities, can form surprisingly similar-looking dunes. This would have radically changed, at

18. Cross-bedded sandstone formed by the migration (from right to left) of wind-blown desert dunes in the Jurassic Period: the Navajo Sandstone at Angel's Landing Trail, Zion National Park, Utah, USA. The cliff face is several metres high.

a stroke, the landscape conjured up in people's minds by those Utah cliffs.

Closer study was made of the Navajo rocks, focusing on criteria that might distinguish wind-laid from water-laid sediments. Wind-blown sands, for instance, contain few mud particles or mica flakes, as these are efficiently winnowed out by wind, while they can more easily settle out in a moment of slack water. And, if the deposits were tidal, they might be expected to show traces left by the tides' ebb and flow, and of regularly changing strength of tidal currents between spring and neap tides. Restudy showed that the characters of the Navajo Sandstone better matched wind than water, so it remains as a classic example of a wind-blown desert sandstone.

Thus, while the rock strata are an enduring body of evidence, their interpretation always remains subject to change. It is harder to interpret environments that are difficult for us air-breathing humans to visit, like the bottom of the sea. This is something of a problem, as most ancient strata on Earth formed in the sea, because—in contrast to our familiar landscapes—that is where sediments are most likely to accumulate and least prone to being eroded away. Geologists have therefore put a great deal of effort into understanding the processes that occur on the modern sea floor, by visiting this (to us) alien environment in submersibles, and trying to recreate submarine conditions in the laboratory. Or, when money is short, and the need is great, in the equivalent of an improvised bathtub. This is what the Dutch geologist Philip Kuenen did in the 1930s, pouring a slurry of sand and mud down the bottom of an inclined water-filled tank. The mixture transformed into a home-made *turbidity current*, enormously larger representatives of which form a highly efficient means of carrying sediment hundreds or even thousands of kilometres across a sea floor, before it settles as a distinctive layer in which the coarse particles settle at the bottom and the finer particles at the top. Fossil examples of such *turbidites* make up a very large proportion of strata formed in deep water (Figure 19).

The problem of interpretation is particularly acute in trying to work out ancient environments from strata on other planets, as planetary scientists now try to puzzle through the pictures of sediments and strata sent back from Mars and from Titan.

Some strata, though, allow insights into environments that we cannot visit on Earth—not even in the best-designed vessel. The pyroclastic currents of incandescent ash and rock debris that speed down the flanks of volcanoes are utterly unapproachable, and are also enveloped in billowing dark clouds that completely obscure their inner workings—even when they are observed from a safe (that is, a very great) distance. However, the strata that they leave behind, once they have cooled, can safely be picked through and analysed to glean useful information about the behaviour of these most fearsome of volcanic phenomena.

19. Regularly layered strata of the kind formed by turbidity currents on a deep sea floor, Norway.

The strata have first to be preserved so they can be studied. In the long term, the only way of effectively preserving them is burial by more sediment layers. Commonly this happens because the local area of crust was subsiding, in effect forming a depression in the ground that is then filled with sediment. On Earth, patterns of subsidence (where strata accumulate) and uplift (where rocks are eroded) of strata are largely mediated by plate tectonics, most spectacularly by the formation of high mountains and deep ocean trenches, but more generally through patterns of crustal warping that take place around these zones.

Sedimentary strata do not only yield information on physical processes. They can be revealing about chemical environments, like the salt deposits that show the former presence of concentrated brines in drying seas. They are also the kind of rock in which the mortal remains of long-dead animals and plants can be preserved, and which form the fossils that palaeontologists study.

Fossils

The type specimen—the one that officially symbolizes the species—of *Tyrannosaurus rex* can be seen in the Carnegie Museum of Natural History in Pittsburgh. It had a picaresque history in keeping with this most melodramatic of dinosaurs, being discovered in 1902 in the aptly named Hell Creek rock formation in Montana by the legendary fossil collector Barnum Brown (named after the outrageous showman Phineas T. Barnum and known to many as just 'Mr Bones', Barnum Brown was immortalized in the title of his second wife's memoirs *I Married a Dinosaur*). The *T. rex* skeleton first went to the American Museum of Natural History, but this institution then sold it to the Carnegie Museum in 1941 for just $7,000, something that the latter described later as not just the steal of the century, but the steal of the last sixty-six million years.

T. rex in turn symbolizes fossils to many people, as rare, dramatic, and fierce—and to be fair to that species, it was all of these, with the largest calculated bite force of any land animal known, living or dead. But it certainly does not symbolize all fossils, which mostly are small—indeed, most are invisible to the naked eye—and represent normal, humdrum, inoffensive living organisms such as snails, clams, and corals that can be extraordinarily common. The living organisms themselves may be seen in all their glory, by scuba-diving atop a modern reef. Drill down through the living skin of such a reef, though, and one will encounter a limestone rock made of the packed-together skeletons of countless earlier generations of these organisms, that may be many metres—or even a few kilometres—thick. Ancient limestones of this kind, representing long-vanished reefs, make up a significant part of our landscapes. There is a small, lovely example near the charming Shropshire village of Much Wenlock, in England, that dates back to the Silurian Period, about 430 million years ago (Figure 20),

20. A piece of Wenlock Limestone—a rock formed as part of a reef dating back to the Silurian Period *c.*430 million years ago, with fossils of reef organisms.

and there is a spectacular petrified reef in the Canning district of north-west Australia which is a little younger, from about 380 million years ago in the Devonian Period.

Abundant fossils extend well beyond coral reef regions. Beach sands may be full of shells, lake deposits can include the remains of fish and washed-in leaves and insects, and deep-sea deposits can include countless sunken plankton and sharks' teeth. Most sedimentary strata include some kinds of fossils—including volcanic ashes, within which some exquisite fossils have been found. These are the kinds of remains that Buffon, Cuvier, and Darwin saw abundantly on their travels, and from which they gained a sense of the way in which dynasties of different life-forms appeared and disappeared on this planet. And since the times of these pioneers, fossils have been seen to be even more abundant in their microscopic forms—a single gram of mudstone can contain thousands of fossilized spores or pollen grains, and a single gram of chalk will contain even greater numbers of tiny skeletons of planktonic algae.

There are a number of themes running through modern palaeontology. One has been to try to link, ever more closely, the fossils to the nature of past life itself. One major problem here has been that almost all fossils are skeletons of one kind or another—shells and bones made of calcium carbonate or calcium phosphate, or the tough organic coatings of pollen grains. Soft tissues such as those of skin, muscles, and gut are fossilized only under exceptional circumstances, and the many kinds of animals that are entirely soft-bodied, such as jellyfish, have an exceedingly poor fossil record.

But when they *do* fossilize, a treasure trove of information can be unearthed. Some of the most celebrated fossil localities are *lagerstätten* (from the German word for 'storage place'; the singular is *lagerstätte*), where particular conditions have led to exceptional fossil preservation of soft tissues, sometimes even down to the cellular level. One celebrated example, in rocks in British Columbia more than 500 million years old, is the Burgess Shale, discovered on the steep slopes of Mount Burgess more than a century ago by the geologist Charles Doolittle Walcott. After he discovered it by chance, for several years he and his wife and children set up camp each summer to hammer—and dynamite—out many thousands of beautiful fossilized trilobites (extinct arthropods), complete with their delicate legs and antennae, worms, soft-shelled crustacean-like animals, and a host of other creatures. This new menagerie transformed (and with continued study, is still transforming) our view of the nature and diversity of some of the earliest communities of animals to have appeared on Earth.

Another fine example is set atop a high plateau west of Recife in Brazil. More than a little reminiscent of Arthur Conan Doyle's *The Lost World*, the setting is home to the Santana Formation, a fossilized lagoon floor from some 110 million years ago in the Cretaceous Period, in which the entombed fish show exquisitely preserved guts and muscles—and even bacteria petrified in the act of feasting on those tissues.

Biological time

While the use of fossils profoundly deepens our understanding of how life evolved on this planet, most palaeontological work is done not so much for its own sake, but to support the framework of the rest of geology. In this task, that biological evolution of many successive life-forms is exploited to provide a fossil-based chronometer for the Earth's strata. This has taken the insights of William Smith in the early 18th century, in seeing that strata of different ages contained different fossils, and elaborated them enormously in modern biostratigraphy, which remains the most routinely used method of rock-dating for geologists.

The basis of biostratigraphy is simple. Each species originates by evolving from some previous species, lives somewhere on Earth, typically for something like a million or a few million years, and then becomes extinct, either through some environmental calamity, or because it is outcompeted by some new species. The fossilized remains of that species, once found in the rock, are a marker for that particular time interval. Therefore, there are potentially many thousands (even millions) of geological time intervals to be recognized by that means. In practice, there are complications.

One complication is simply the sheer number of different types of fossil, which reflect the diversity of life, even among that fraction of life that possesses fossilizable skeletons. Most major fossil groups, such as corals, ammonites, and trilobites, now include many thousands of individual species (and more new species continue to be recognized every year). Furthermore, the shells or carapaces of those organisms can be squashed, or broken, or only partly preserved in the rock, and so it is a real skill to recognize particular species in this way. That skill is built up only slowly by the specialist palaeontologists involved, many of whom over a lifetime's work only focus on the fossils of one major group—thus, there are ammonite specialists, trilobite specialists, coral specialists, fossil pollen specialists, and so on. Even these specialists tend not to set up separate time units based upon individual species, but lump them together so that communities of organisms that lived at the same time form the basis of fossil-based time units, or *biozones*, that underpin the modern Geological Time Scale.

With this division of labour, remarkable precision of time division can be attained. The twenty-five million years of the Silurian Period, for example, are divided into some forty or so successive biozones on the basis of graptolites (extinct ocean-going plankton) (Figure 21). These biozones are only effective in strata formed in the deep sea, though, so to *correlate* into time-equivalent strata formed in shallow seas or on land, other biozones based on other Silurian fossils must be used, such as those based on brachiopods ('lamp-shells') or fossil plant spores.

Some fossils are more useful than others. The bones of *Tyrannosaurus rex*, magnificent as they are, are not terribly useful for biostratigraphy, being large, rare, fragile, and confined to part of north America. The really useful fossils are small, common, widespread, and rapidly evolving, like the graptolites of the Silurian or the ammonites of Jurassic and Cretaceous times. The microfossils are particularly useful, even though specialized and somewhat scary techniques have to be used to extract them from the rock, such as the use of powerful acids. However, as thousands of microfossils may be extracted from a small chip of rock, they can readily be obtained from the pounded-up 'cuttings' obtained when borcholcs arc drilled, a process in which larger fossils are obliterated.

21. A rock surface with examples of the fossil graptolite *Normalograptus persculptus* (each reaching about 2 millimetres wide). This species of extinct plankton represents the very end of the Ordovician Period, about 444 million years ago.

There are now sets of biozones for each geological time period back to the beginning of the Cambrian Period, 541 million years ago, when easily visible fossils first became common. Their use extends to the present day, where the artefacts made by humans can be used as 'technofossils' to date very recent strata—such as in the precise dating of coastal cyclone deposits in Asia via the date stamps on the food-packaging litter that they contained.

The detailed subdivision of geological time by fossils has brought about enormous refinement of the Geological Time Scale. But there are other means of subdividing time in sedimentary rocks, such as by tracing changes in their chemical properties. These chemical changes in turn often relate to changes in climate or in environmental conditions on Earth—some as regular cycles, and other events that were haphazard, sudden, and catastrophic.

By means of such interconnections, the timescale is evolving into an integrated history of the Earth.

Climate connections

Among the most widely used fossils contained in strata of the last hundred million years are the *foraminifera*. These are single-celled, marine, amoeba-like animals, which secrete elegant shells of calcium carbonate that they live in, extending pseudopodia into the water to feed on yet smaller organisms. Some are planktonic, living in the surface layers of the ocean (Figure 22), and others live in the deep sea floor muds. Like most organisms they are sensitive to temperature, and each species has its own tolerances, some preferring warm water and others liking cooler conditions. If a palaeontologist knows these tolerances, they can use such

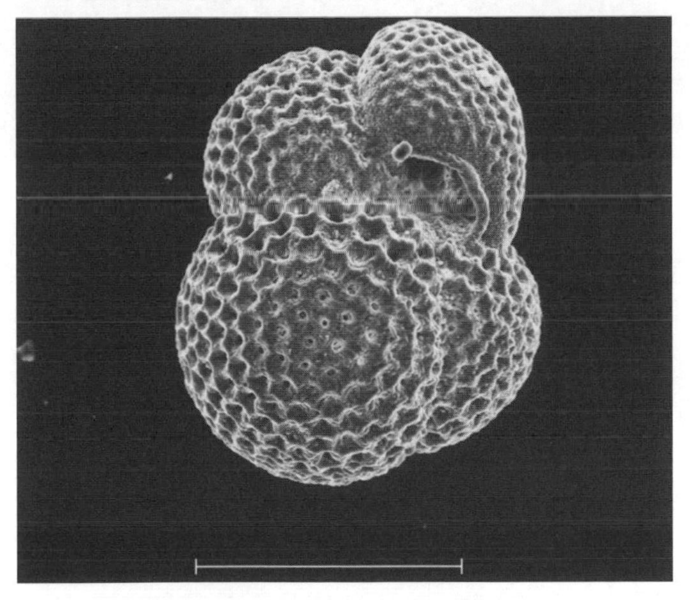

22. Scanning electron image of the microscopic skeleton of a planktonic foraminifer. The scale bar is one-fifth of a millimetre.

microfossils to assess not only the age of the strata, but also the temperature of the water in which they lived, and therefore gain an idea of past climate.

The foraminifera, though, contain more climate signals, locked away within their tiny skeletons—they can reveal how much ice was present in the world when they were alive, even if they lived thousands of miles from the nearest icecap. The ingredients for these skeletons are taken from the waters around them, including the water molecules themselves. Within the water (H_2O) molecules, the oxygen is represented by two stable (i.e. non-radioactive) *isotopes*: 'normal' oxygen with eight neutrons and eight protons, or ^{16}O, and small amounts of a heavier isotope with ten neutrons and eight protons, ^{18}O. Water molecules with ^{18}O are heavier and evaporate less easily, and so the clouds that drift from the oceans, some eventually to fall as snow on to the icecaps, are enriched in the light, ^{16}O, isotope. If icecaps grow at the expense of ocean water as climate cools, they will draw ^{16}O out of the oceans too—and the foraminifera will record this in their shells as a relative increase in the ^{18}O isotope. When they melt as climate warms, the ^{16}O comes flooding back—and the new generations of foraminifera record this change too. Their skeletons, falling into the oozes on the ocean floor, form a magnificent, detailed archive of climate change, which has been one of the main targets of the Ocean Drilling Program (see Chapter 3).

The record of climate may be read in many different kinds of strata. The record of strata on land tends to be patchier than that of the oceans, because on land much erosion takes place, and so there tend to be many gaps in the strata preserved at any one place. Nevertheless, the evidence that remains can be strikingly detailed, and there are some places—such as deep lakes and long-lived peat bogs—that can preserve a continuous climate record over many thousands of years. One of the best of such climate archives is within the rock known as ice, in the form of layers of snow and ice that have built up to form the icecaps of

Greenland and Antarctica. These contain few or no fossils, but the snow layers preserve a record of temperature in the form of those oxygen isotopes, and when the snow layers are deeply enough buried to compact down to form ice, this ice includes many air bubbles that can be analysed to show how much of the greenhouse gases, carbon dioxide and methane, were in the air during times of warm and cold climate of the Ice Ages. The ice layers also contain wind-blown dust (air is dustier in cold, dry, glacial climates) and yet other signals, such as acid layers from distant volcanic eruptions. The ice-sheets can be sampled by drilling boreholes through them, through ice that is up to 4 kilometres thick and that may extend back to a million years ago (800,000 years-worth have already been sampled), on Antarctica.

The clues that have been gathered from strata in the oceans and on land have, in the past few decades, transformed our understanding of how climate has changed in the geological past. There have been marked changes between warm and cold climate that have been tightly linked—at least in the last 800,000 years, as we know from the ice core record—with rises and falls in greenhouse gases. Both climate and greenhouse gases show a strongly regular, cyclical pattern, of intersecting cycles of 20,000-year, 40,000-year, and 100,000-year durations. These cycles are essentially astronomical, as predicted by earlier scientists, notably the Serbian mathematician Milutin Milankovitch in the early 20th century. They represent the periodicities of change in the Earth's 'wobble' around its axis, in the angle of the Earth's spin axis, and in the shape of the Earth's orbit around the Sun. These *Milankovitch cycles* were termed the 'pacemaker of the Ice Ages' by the scientists Jim Hays, John Imbrie, and Nick Shackleton, who used ocean floor sediment cores to pin down this connection between astronomy and climate, and so to vindicate Milankovitch's hypothesis. The very small changes in the amount of seasonal pattern of sunlight, amplified by rises and falls in greenhouse gases, led to the regular changes between warm and cold of the Ice Ages.

Strata from a hundred million years ago show that the Earth was much warmer then, with little or no ice on Earth—as shown by fossilized trees and dinosaur bones recovered from Antarctica and the Arctic Circle. Even then, the Milankovitch cycles influenced the Earth—they can be seen as metre-scale stripes in the chalk strata, reflecting changes between wetter and drier conditions. Whether in the greenhouse or icehouse times of our planet, astronomy has been a constant orchestrator of environmental conditions, of metronomic regularity. Other changes, though, could come out of the blue, to make their mark in the rocks.

Catastrophe strata

The pale limestone strata in the steep valley of the Apennines at the small town of Gubbio, in Italy, do not look, at first glance, to record any dramatic event. However, if successive samples are hammered out of them and patiently examined by microscope, they show the outlines of fossil foraminifera. The bottom layers show a range of large foraminifer species, up to 1 millimetre across. Then, directly above a thin band of red clay, less than 1 centimetre thick, there is a sudden change. Most of the foraminifera species disappear—and never reappear, being replaced by a very few, diminutive forms. Clearly, something happened to quickly, thoroughly, re-fashion the nature of the ocean plankton.

This level was identified as marking the change from the topmost Cretaceous strata, some sixty-six million years old, to the earliest Tertiary (now termed Paleogene) strata. The level seemed to match up—as best as could be worked out—with the level where the (non-avian) dinosaurs suddenly disappeared, all over the world, and, in other marine strata, where successful, long-lived animal groups such as the ammonites also died out. This great dying had long been one of the great mysteries of Earth history. But what caused it? Many ideas had been put forward, from climate change to evolutionary self-destruction to plagues of insect pests that laid waste to the vegetation the animals depended on.

A geologist, Walter Alvarez, and his father Luis, who happened to be a Nobel Prize-winning physicist, began to explore the mystery. When they came to Gubbio, they simply wanted some guide to how quickly the layer of red clay at the boundary accumulated. If it accumulated only very slowly, it would make the apparent sudden extinction of the Cretaceous foraminifera more of a long-drawn-out event. They thought a good time marker might be the chemical element iridium. This, an element related to platinum, is rare in the Earth's crust, but arrives to Earth from outer space, slowly but more or less steadily, as meteorite dust. If the red layer contained a little more iridium than the limestone, this might indicate its slow accumulation.

They analysed across the boundary, and found huge amounts of iridium in the red clay layer—thirty times more than the tiny levels in the limestone, much more than they would expect even for an exceedingly slowly accumulated layer. Furthermore, other, thinner, red clay layers above and below did not show any increases in iridium content. Some event had clearly happened at the exact level where the fossils suddenly changed (Figure 23).

The Alvarez team in 1980 suggested the iridium had come from a massive meteorite impact on Earth, the effects of which would have precipitated ecological collapse and the death of many species. It was a highly controversial suggestion, not least because the influence of Charles Lyell and of his gradualist philosophy was still strong, a century after his death, and any catastrophist explanations were fiercely resisted.

The same pattern was soon found elsewhere in the world at the same level. More evidence of impact was found, too, including tiny fused glass spherules and grains of shocked quartz. The crater itself was later recognized, an enormous buried structure about

23. Luis and Walter Alvarez at the Cretaceous-Paleogene boundary at Gubbio, Italy.

180 kilometres across, in Mexico. There is still debate on the nature of the kill mechanisms, and of the compounding effect of other perturbations at that time, such as massive volcanic eruptions. Nevertheless, it is now clear that the Earth's evolution has included both regular and gradual influences and also

episodes of sudden massive change from more or less random events. Both Cuvier and Lyell can be said to be partly vindicated by the work done since their day.

The evidence of geology is spread across the entire world, and therefore one needs to explore that world to find it. Geological fieldwork is the perfect way to do this.

Chapter 6
Geological fieldwork

Geology surrounds us constantly—and useful, absorbing, practical geological study of one form or another can be carried out pretty well everywhere. True, our museums and research laboratories include rare and exotic mineral specimens, and the bones of mammoths and dinosaurs, together with sophisticated analytical equipment to analyse these minutely. But, among the many wonderful things about geology is that one can walk out of the back door, armed with little more than a sense of curiosity, some background knowledge, and a small magnifying lens, and still make significant discoveries, even today. The more one indulges in such small adventures, the more one gets drawn into a fascination with the infinite possibilities of a science that has many millions of years of crowded planetary history as its field of play. This sense of wonder and curiosity is basically what drove the beginnings of the science, in the days of Buffon and Hutton, Mary Anning, and William Smith. It is still what motivates geologists today, very few of whom hang up their boots for good when they retire. And it is in geological fieldwork that the practical outlet of this curiosity finds its purest expression.

The explorers

The rare, exotic, and wonderful has, of course, been one of the great attractions of the science, often being the 'hook' that initially attracted the attention of many a geological enthusiast—and of many sober professional geologists. This kind of attraction dates back at least from Greek and Roman times (and probably well before then, although no records exist) when bones of mammoths and dinosaurs, now and then accidentally unearthed, were seen as impressive proof that giants, ogres, and monsters really existed. Ancient travellers' tales of a supernatural beast, the griffon, too, have been plausibly linked to some of the dinosaur bones (specifically, of *Protoceratops*, an early relative of the more famous *Triceratops*) that litter parts of the landscape of the far-off mountains of Mongolia; the travellers, seeing those finely preserved bones, likely imagined that the living creature could await them around the next hill. This fossil-rich region became a focus of some of the most famous expeditions in more recent times.

Roy Chapman Andrews is said to be one of the real-life inspirations for the adventure film character Indiana Jones: he was scientist, explorer, dinosaur hunter, and a crack shot who fought with brigands. As a young man he wanted so much to work at the American Museum of Natural History that he took a job there as a janitor; he went on to become its director. In the 1920s he led several expeditions to Mongolia's Gobi Desert, needing to negotiate the unstable political landscape of the China of those times to get there. They discovered magnificent specimens of gigantic early mammals and of dinosaurs, including the first known nests of dinosaur eggs. These feats made him famous (he was not shy to promote himself, writing many books of his real-life explorer stories as well as scientific monographs) and stimulated great enthusiasm for the science. It is probably appropriate that an impressive mammal carnivore, *Andrewsarchus*, with ferocious teeth set in a jaw half a metre long, was named after him.

Such explorations were made not only for dinosaurs, but targeted precious gems and metal ores, some of which were found accidentally, rather like the fictional Lord John Roxton of Conan Doyle's *The Lost World* (perhaps a role model for Roy Chapman Andrews, one suspects), who noted a very specific kind of blue clay amid the derring-do on the high plateau, and so had gained a pocketful of diamonds by the time the heroes made their escape. The prospectors for precious metals and diamonds who sparked such things as the 19th-century gold rushes in North America were similarly looking out for particular signs and clues in the rock formations for the treasures they were seeking. The more successful—and luckier—prospectors learned to tell the enticing sparkle of fool's gold, or pyrite, from the lustre of the real metal, and became adept at tracing promising-looking quartz veins along the hillsides, just as the successful dinosaur hunters had an eye for the right kind of strata and the specific patterns that a few tiny bone fragments can make when scattered among wide expanses of rock scree.

Today, many people carry out geological fieldwork of this kind, as an absorbing pastime, and there are many kinds of help available to people who wish to do this, from geological field guides to different areas, to groups and communities of geologists, both professional and amateur (such as the Geologists' Association in Britain) who organize field trips and field meetings in convivial settings.

This kind of prospecting, though, is not the same as systematically working out the total geology of a landscape—what kind of rocks compose it, how they were assembled in geological time, and what they mean in terms of the entire history of that landscape. Here we have to go back to the approach of some other of the geological pioneers, and see how their insights were developed into what we understand today as geological mapping.

A landscape in four dimensions

The aristocratic Comte de Buffon, working incessantly as a kind of all-purpose savant in the 18th century, was not an adventurer in the mould of Roy Chapman Andrews, or of Humboldt in the 19th century, whose exploration of South America showed similar daring. Buffon's life revolved around Paris in the winter and his country estate in the Bourgogne district of France in the summer. Nevertheless, he developed a sharp eye for his local countryside, seeing where shales had been dug into on the valley floors, while there were limestone crags higher upslope. In explaining these as a three-dimensional layercake of two units of strata of different age, with older shales below and younger limestones above, he was in effect creating a four-dimensional interpretation of the landscape, in which puzzling out the physical structures of the rock masses that underlie a landscape led directly to working out the history of how the rocks formed.

Resolving this geometrical puzzle was just one part of Buffon's many-sided investigations into the Earth and its living organisms. But this kind of geological analysis became the obsession of William Smith ('Strata' Smith became his nickname), an artisan from a humble background. From the late 18th century, Smith worked as a land surveyor to help the building of the growing network of roads, canals, and railways, and also inspected coalmines, then burgeoning as the power source for the Industrial Revolution. He observed that the strata formed successions of different rock types that were coherent and *predictable*, particularly when combined with the use of fossils to help characterize the units he recognized (a self-taught skill at which he became adept).

Smith could therefore trace a stratal unit around the countryside as an *outcrop*, an area where the geological unit intersects with the ground surface. The outcrop could be shown as a coloured area on a map, separated off from the outcrops of the rock units above and below (which would be shown in different colours) by lines on the map. Thus, he could convert a standard topographic map by this means into a *geological map* (Figure 24). William Smith took his obsession to heroic levels, and over his lifetime mapped out the

24. Part of the geological map published by William Smith in 1816. The base of each rock formation is shown as a darker shade, to make the map clearer.

geology of much of Britain. This was geological fieldwork across a whole country, travelling by horse-drawn carriage and on foot and, incidentally, going bankrupt in the process. The map, now with pride of place in Burlington House, the home of the Geological Society of London, is a work of genius as well as of obsession.

The map does not show the end of Smith's thinking—not least as it is only a two-dimensional sheet of paper. To display the third dimension, Smith systematically constructed *geological sections*, imaginary vertical cliffs to show how the strata are arranged underground. Smith had not been underground to check this out. The sections he drew were projections, of where he predicted the rock units *should* be. If he could see that, at the surface, the rock strata showed a certain angle of tilt in a certain direction, he would project them underground at that angle. In some sections he also projected the rocks into the air, as a kind of 'ghost strata', to show where they once had been, before they were eroded away, their detritus washed away in times long past to become new strata elsewhere.

These were not only projections, but practical predictions. Take the valuable coal strata that he had taken an interest in. Before the kind of insights he prompted, many speculators in the coal trade had invested their money by drilling or digging underground more or less at random. If they found coal, they became rich. If not, they lost their money. Smith's method predicted those places where there was reasonable chance of finding coal (and would even give an indication of how deep that coal might lie). Similarly, he could indicate those places with little or no chance of finding coal, and where any investment would be squandered.

The fourth dimension was a natural outcome of these constructions. The lower strata in such a layercake of rock were older, and the ones above were younger. And, after they were laid down, they were, over a long period of time, uplifted, tilted, and eroded—a process still taking place.

The bare bones of William Smith's work seem straightforward, even obvious. But they are only obvious when looking at the results. The evidence he had to go on was not so simple at all.

Assembling the evidence

What Smith had to cope with all his life is what every geology student feels when let loose into fieldwork for the first time. A geological map is a mass of coloured stripes and patches that show how the different rocks are distributed across an area. But go to that area, and what does one see? Green fields, meadows, woodland, and towns and villages—with precious little rock that is obvious, if any at all. Approaching more closely, the student might push aside the vegetation to see what lies below—to find only a thick soil layer. The telltale rocks must lie somewhere below—but where?—and how can one find out what they are?

These are the kinds of obstacles a geologist must overcome to understand the geology of a region. Not everywhere, for sure. In arid regions of the world, there is little vegetation, soils are thin, and rocks are widely exposed at the surface. In such regions the geology of an area may be clear at a glance, particularly if the rocks are simple and uncomplicated—one might think of the eloquent landscape of the Grand Canyon in Arizona. But in most parts of the world, the geological bones of the landscape are thickly clothed in soil, subsoil, and vegetation, and now increasingly also by the carapace of our urban constructions. What, then, is the geologist to do?

Geological fieldwork, and particularly the systematic geological mapping of a landscape, is par excellence the kind of study where highly fragmentary and diverse information is gathered and integrated to try to produce a sensible working model of the geological structure of an area. The model is exactly that—a model—which has to fit the evidence as well as possible, and is always open to being revised when new evidence comes in. It is best regarded as a kind of puzzle, where only a few of the necessary clues are available, but which nevertheless has to be solved somehow, even if only in an interim fashion. It is not an exercise for those who need certainty in their working life. But, for those who adjust to its uncertainties, it offers an almost infinite fascination—and one, moreover, that can be indulged out in the fresh air. Where, though, does one start?

One way to begin is to look for rock *exposures* (an exposure is different from an outcrop: there is outcrop everywhere of some kind of rock—but that rock may be only very rarely *exposed* at the ground surface, to be available for examination). Rock exposures may need a little hunting down. There may be small quarries in farmyards, or a little bedrock might be visible in the bottom of a stream or river, or in a roadcut. In very poorly exposed ground, geologists can even resort to seeing what kind of rock fragments have been brought up by rabbits digging burrows! One thing that helped William Smith were the artificial excavations made through bedrock as roads and canals were constructed. But these exposures are still only a tiny proportion of the land surface, often much less than 1 per cent. Nevertheless, the rocks seen in these can be put into a pattern relative to the rocks in other exposures, and if the strata are tilted, the direction of tilt can be used to predict where there are younger strata (above) and older strata (below). It is the beginning of a solution for the puzzle.

Another factor helps—the shape of the landscape itself. Some rock units are hard, and resistant to weathering. These will stand out as ridges of higher ground. Softer rock units will form lower-lying ground, and the boundary between hard and soft rock units can often be seen as the change from a steep slope to a gentle one. One example that was very familiar, and useful, to Smith was the chalk strata of southern England, which forms lines of hills (the North and South Downs, and the Chilterns). Beneath the chalk, there is commonly a thick unit of soft clay, which forms a swathe of low flat ground. Smith would certainly have used the easily visible

25. Lines of ridges (scarp slopes) in the landscape showing where hard-weathering strata come to the surface. Tracing out such topographic features allows the geological structure of a landscape to be worked out.

division between these two types of topography to trace out the boundary between these two major rock units. Exploiting the way that geology influences topography was one of the major reasons why Smith could cover such an enormous area of ground single-handed; he became, in effect, a highly skilled landscape psychologist. This technique is still a mainstay of geologists, and can be used to a very fine degree. A hard sandstone bed only a few centimetres thick set within soft shales might produce a ridge so slight at the surface that a geologist needs to get down on hands and knees to see it. Undignified, perhaps—but if the ridge is at all perceptible, then it can be used to help trace the geology (Figure 25).

Not all geology follows such more or less regular patterns. Where William Smith came to the ancient mountains of Wales and

Scotland these became in effect a 'here be dragons' part of his map, because these strata have been thrown into folds and torn apart along geological faults by mountain-building movements. This kind of terrain was too complex for Smith's pioneering broad-brush work. The geology of such landscapes can be teased out, but here the geologist must move slowly and carefully to decipher such deformed and metamorphosed rocks. In this kind of terrain also, large masses of magma often invaded the rock mass, pushing it aside to cool and solidify as large granite masses and other such bodies of rock that cut across the original strata. Remarkably, one example was beautifully analysed by James Hutton, in the late 18th century. On the lovely Isle of Arran off western Scotland, Hutton drew a geological cross-section of the mass of granite of that island pushing upwards from below and shouldering aside the strata on either side; it is an interpretation that, in its essentials, has stood the test of time.

The drift deposits

Between the ancient rocks and the modern vegetation and soil at the surface, there is generally another deposit, which can be tens or even hundreds of metres thick. This material greatly puzzled the early geologists: often completely unlike the ancient rocks beneath and sharply separated from them, it includes chaotic masses of boulder-rich clays and spreads of gravel. These deposits, once thought to be the relics of a biblical Deluge, are now revealed as deposits formed by glaciers and streaming meltwater during the recent Ice Age.

These are still widely called 'drift' deposits (from the old idea that many formed from drifting icebergs), but are more technically referred to as 'superficial' or 'surficial' deposits. Not all relate to glaciation, by any means. In low latitude parts of the world they can include thick windblown sand deposits, as in parts of the Sahara, or the thick windblown silt or *loess* that covers much of central China, having been blown there during the past two and a half million years from the Himalayas, and peat bog deposits. Where large rivers such as the Mississippi or Ganges-Brahmaputra meet the sea, huge deltas build out.

'Drift' deposits drape over the ancient rock strata, usually with an enormous time gap—an unconformity—between the two juxtaposed types of geology. As with the ancient rocks, they can form characteristic forms of topography. Active present-day rivers build the alluvial deposits that underlie river floodplains. These may be flanked by step-like patterns of 'river terraces'—deposits of older generations of floodplains, through which the modern river has cut.

Among glacial deposits, spreads of 'boulder clay' or glacial till are not usually so conveniently regular, as the moving ice simply smeared this material across the underlying ground. Nevertheless, they might be recognized by subtle changes in slope, or by different soil or vegetation types (Figure 26). Some kinds of glacial deposits form specific topographic features, such as the elongated dome-shaped *drumlins* that formed in places beneath moving ice, or the barricade-like *terminal moraines* that may stretch across valleys, marking where a glacier was at a standstill in its retreat, and piled up rock and sediment debris like at the end of a conveyor belt.

A new and rapidly growing addition to the 'drift' category is represented by the deposits made by humans: embankments, landfill sites, landscaped ground of various kinds, shot through with cavities, pipes, and tunnels. In and around urban areas such layers of 'artificial ground' can be several metres thick and cover many square kilometres. Their diverse and unpredictable nature makes for a distinctive new addition to our planet's geology.

All these rocks and deposits may be approached simply by the traditional means of the geologist, with hammer and hand lens and a keen eye for the landscape. But the new technology can be useful too.

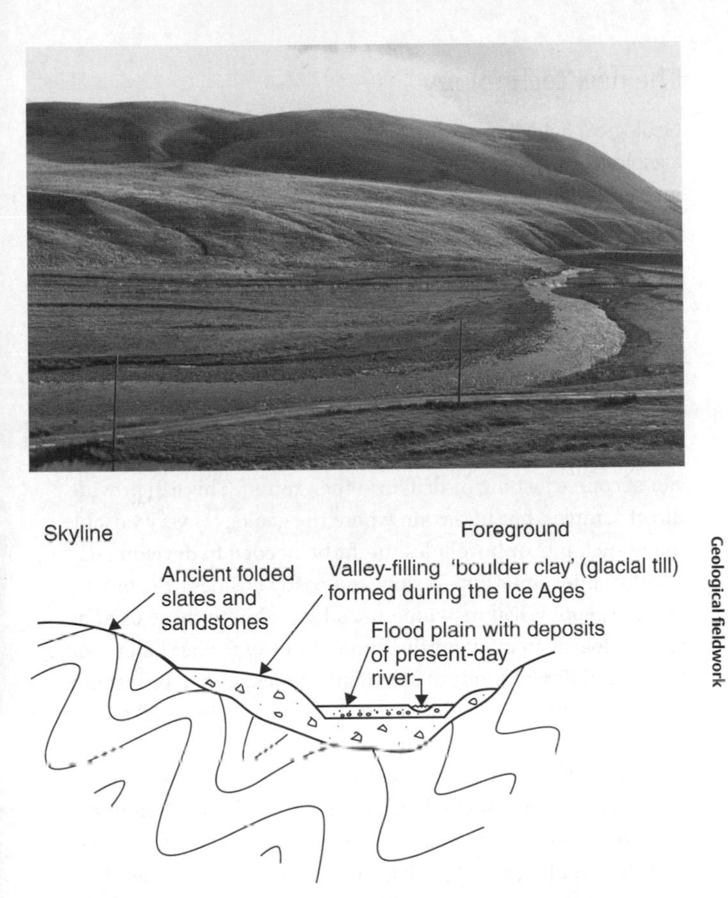

26. A Welsh landscape dominated by superficial 'drift' deposits. The foreground is made up of the modern river floodplain underlain by alluvial deposits, which are still forming today. The middle ground with paler vegetation is underlain by glacial deposits dating back about 20,000 years. Only the hills on the skyline represent ancient geology: beneath the grass and thin soils are 400-million-year-old Silurian slates. The sketch cross-section below shows the nature of the geology below the ground surface. Developing the ability to 'read' such landscapes is part of geology.

The new technology

Geological fieldwork is supremely the art of the possible, and a geologist will always be able to provide *some* geological interpretation of an area, even if it is only of the 'here be dragons' kind of categorization that William Smith gave to areas too complex for him to interpret with the little time and very few resources he had available. But, in recent decades, ingenious new means, mostly technology-based in one way or another, have been found to glean information even from the most unpromising terrain.

If no useful geological clues are seen at the surface, then one can of course just dig or drill into the ground. This can provide direct samples, but in terrain where the geology is very variable, many such pits or boreholes might be needed to develop an effective interpretation—a slow and costly business. So, one can use geophysical instruments, a little akin to those used by archaeologists to map out the foundations of former habitations, to detect different kinds of sediment or rock. Clays, for instance, are quite electrically conductive while sands are electrically resistive, so with a portable ground conductivity meter a geologist might quickly work out the precise distribution of high conductivity (clay-rich) versus low conductivity (sand-rich) areas. Ground radar devices can be towed across the ground, generating images that show underground geological structures, while other devices can measure magnetic properties, or natural radioactivity, or even minute gravitational differences between different kinds of rock. Not every technique works everywhere, of course—but where they do work well a geologist can feel that they have X-ray eyes, and can freely explore an underground realm that was previously frustratingly opaque.

As technology evolves, newer techniques arrive to bolster the geologist's armoury. Satellite images can be used to detect different types of soil or vegetation through changes in the light spectra they create, which might betray different kinds of underlying geology. Airborne laser detection techniques (such as 'LIDAR') can detect tiny changes in topographic height, and these also have proved effective in detecting such things as former river-courses. And, the maps and notebooks of geologists are now being supplemented by portable computers, in which data can be manipulated to produce three-dimensional images and interpretations while the geologist is still in the field.

Field geology, thus, is evolving and adapting to new conditions and new possibilities. Much geological fieldwork is done for practical reasons quite beyond the sheer interest in working out the geological history of an area. So, the many ways in which geology underpins our society and economy is the next topic to explore.

Chapter 7
Geology for resources

Practical beginnings

Humans have been practical geologists since even before our own species, *Homo sapiens*, walked the Earth (Figure 27). In Ethiopia, there is a sandstone massif, the Messak Settafet, that was worked continuously for half a million years, leaving a landscape which is still littered with stone tools and worked rock debris. Five thousand years ago, Neolithic peoples dug down through more than 10 metres of chalk strata at places such as Spiennes in Belgium and Grimes Graves in southern England to reach particular layers of flint, specifically targeted as they provided the most prized raw material for polished, razor-sharp axe-heads. Four thousand years ago, Bronze Age miners, armed with little more than 'hammer stones' and antlers, dug down through up to 20 metres of solid rock at Anglesey's Parys Mountain, off the coast of Wales, to extract copper ore.

The exploitation of such rock-bound resources required a level of understanding of rock composition and structure that remains impressive today. The knowledge would have been passed down from generation to generation and from region to region. The feats of the ancient Parys Mountain miners are particularly astonishing—the complex structure of those copper ores can bemuse a professional geologist today—and were likely inspired

27. An arrowhead made from a flint-like rock, fashioned in the north African desert over 5,000 years ago: an early, and sophisticated, use of geology.

by travellers who had seen copper mining already long-established around the Mediterranean. Such practical skills evolved and became increasingly sophisticated as human civilization grew and developed.

These early geological skills were needed every day, and they always had to adapt to specific local circumstances. Locating and then quarrying the Welsh basalt monoliths that were transported more than 100 kilometres in the construction of Stonehenge required a different specific understanding of practical geology than did the surveying and hewing of the enormous blocks of limestone that built the Egyptian pyramids. The skills were not just needed for displays of royal power. The underground tracing of tectonically contorted seams of slate for the roofing of cottages, or the location of underground salt layers to preserve food over winter, or the mining of coal by the Romans, required similar levels of geological insight—even if a formal science of geology was not to arise for many centuries.

The mining of metals, the extraction of building stone, the engineering of waterways, developed over the centuries. Today, this kind of activity has reached such a level of scale and sophistication that we live surrounded by the products of geology, to a far greater scale than were our hunter-gatherer and early farmer ancestors.

Modern times

The familiar world that we build around ourselves mostly comes from geology, in one way or another. The houses in which we live, and the offices and factories in which we work, are made of reconstructed sand, gravel, mudrock, and limestone—with a few nicely polished slabs of granite or marble thrown in as decoration. Many of these constructions have an internal skeleton of steel, taken from huge iron ore deposits formed near the dawn of our planet, and that steel is also an integral part of the tools we use and the cars we drive along roads that are long strips of crushed rock bound with bitumen. Other metals such as copper, aluminium, zinc, and lead add to those constructions, while yet more unfamiliar ones—such as neodymium, hafnium, and europium—are now crucial to the electronic devices of the computer age. We use immense amounts of energy, mostly taken from coal, oil, and gas in the ground, while part of that oil now goes into the plastics that make up the clothes we wear, the carpets we walk on, and the packaging around the food that we buy at the supermarket. Geological materials are utterly pervasive within our lives. So how do geologists seek those out from within the crust of our planet?

Fossilized sunshine

Coal, oil, and gas are wonderfully convenient energy sources: packed with stored energy, easy to transport—especially the oil and gas, that can be simply pumped through pipelines—and available in very large amounts. In 2016, more than ten billion tons of these fossilized hydrocarbons were burnt around the world, and they account for about 85 per cent of the world's energy consumption. For now, they are indispensable to our lives.

Geologically, all of them share a similar origin—they represent the capture of the sun's energy by plants, which use this energy to convert carbon dioxide from the air into carbohydrates to build their own tissues. Most plants simply decay when they die, releasing carbon dioxide so the cycle can begin anew. But sometimes a fraction of this plant material can be buried deeply within strata, where heat and pressure convert it to fossil hydrocarbons. Coal forms on land, typically by the burial of swamp forests, with gas being released as part of the underground changes that affect buried plant material. The formation of oil starts at sea, with the growth of microscopic marine planktonic algae, which on death sink to the sea floor and are entombed in marine muds; unlike land plants, these plankton are rich in fats and fatty acids, and so their transformation during deep burial releases both oil and gas.

For geologists to locate these fossil hydrocarbons in the rock strata, they need to understand the way in which they formed and transformed in their long subterranean sojourn, and be aware of the circumstances that allowed them to concentrate in rich abundance—and those that dissipated or destroyed them. For the coal forest trees to pile up, generation upon generation, the amount of waterlogging around their roots must be finely balanced: too little water, and the dry aerated conditions speed decay of dead plant matter; too much water, and the ground is too waterlogged for trees to grow at all. The accumulating plant detritus must not be mixed with too much sand and mud from the rivers that flow through the swamp, because that would give a lower-quality coal that produces much ash waste on being burnt. Sometimes a coal seam can just disappear as it is worked—often this is due to a 'wash-out', where a prehistoric river cut through the swamp vegetation to replace it with channel sands, and the seam will usually reappear when the other side of the fossilized river is reached.

These are just the conditions when the plants are growing and dying. There is then the whole subsequent geological history to work out. How many coal seams are there and how thick are they? This is controlled by such factors as the balance between flooding of the forest by the sea (leading to its complete burial by thick layers of sediment) and subsequent re-emergence of the land (so that forests can begin to grow again on the newly exposed landscape). Then, how deeply were the strata buried?—for the more heat and pressure they are subject to, the higher the 'rank' of the coal, as it becomes more carbon-rich when other chemical components are driven off. When the coal strata were uplifted, how steeply were they tilted, and was the whole coalfield broken up by geological faults, so that blocks of strata—together with the coal seams they contain—were displaced upwards or downwards relative to other stratal blocks? The answers to these questions may be worked out by detailed geological mapping of the surface, while boreholes may be drilled to locate coal seams underground (not too many boreholes, though, for drilling is an expensive business). For profitable coalworking, the geologist must think through the entire geological history of the coal deposits, from beginning to end.

Coal seams (Figure 28) at least stay where they formed. Not so oil, which takes a much more tortuous path through the buried rocks. At the beginning of this path, how thickly did the planktonic algae bloom in the seas where they lived? This will depend on such things as climate conditions and nutrient supplies from rivers. Then, what prevented the dead algae on the sea floor from simply decaying and releasing dissolved carbon dioxide back into the seawater? This might be because of too-rapid burial by sediment, or by stagnant, oxygen-deficient conditions at the sea floor—which again is often a result of some climate condition. As the plankton-rich mud is buried to become a *source rock*, at some stage—usually when the temperature rises to between 50 and 150 degrees Celsius—oil and gas will be released. But, once the temperature rises beyond this, then the oil itself is broken down (the rock is said then to have completely passed through the 'oil window') and only gas will be produced,

28. A coal seam, Trevane, Saundersfoot, Wales.

so the geologist must find 'palaeothermometers' that will allow the temperature history of the rocks to be assessed.

The oil and gas then travel upwards through the rock, because they are less dense than water, the other fluid that saturates the strata. The oil and gas travel until they find a rock that is both *porous* (i.e. with many open spaces between the solid rock particles) and *permeable* (in that the pore spaces are large and interconnected enough for fluid to easily flow though the rock). Thus, a coarse-grained sandstone or a fractured limestone can act as a *reservoir rock*—and from that reservoir they would travel on and on too, unless above it there is an impermeable *caprock*, perhaps a layer of clay or salt, and that caprock must be shaped into a *trap*—perhaps by being crumpled tectonically into a dome shape so the hydrocarbons can travel no further (Figure 29), or perhaps because the reservoir rock is an isolated ancient sandy river channel, surrounded by impermeable clay.

It is like a multi-dimensional jigsaw puzzle—or perhaps a fantastical pinball machine—that has been evolving over many

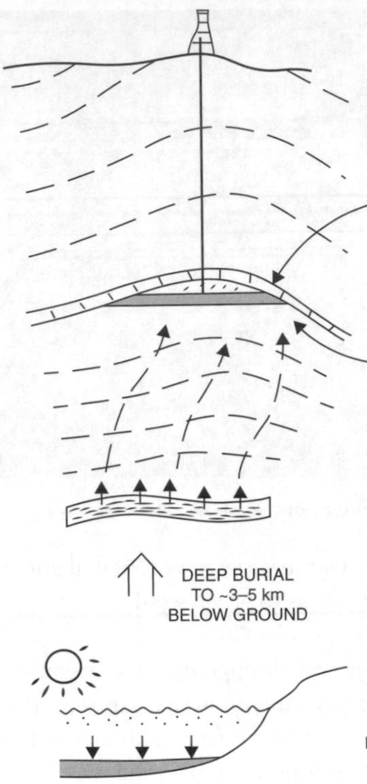

29. How oil and gas is formed.

millions of years. To solve it, the geologist must seek to find and put in place a whole array of clues of different kinds, from the rock types and what kind of ancient environments they represent, to the fossils in the rock that betray their precise age, to geophysical soundings taken to gain images of the three-dimensional shapes of rock bodies, and many more besides. It is now a hugely sophisticated pursuit, and one which continues to evolve, as oil and gas fields are sought in strata beneath the deep sea floor, and as geologists look for gas which is compressed in the pores of fine-grained shales. This 'shale gas' can be released by 'fracking' the shale—forcing liquid in through boreholes at high pressures so that networks of fractures open up, releasing the gas. Shale gas has spawned a whole new field of study and new questions. What controls the gas content in the shales, for instance, and what determines their 'frackability'?—for it is clear that some shales yield much more gas than others. The answers to these evolving questions are among the factors that will determine the future of this enormous global industry.

The metal question

Metals have been used since ancient times, and have defined human cultures, as in the Bronze Age and the Iron Age. Very few metals are present in nature in a more or less pure (or *native*) state, although gold is one exception. Typically, metals have to be extracted from various ores, each of which has formed in specific physical and chemical (and sometimes biological) conditions. A geologist must relate these conditions to particular geological circumstances and settings, and these may show a quite bewildering (or fascinating, depending on your point of view) complexity and variety, which reflects the huge variety of conditions of our planet.

Iron is, occasionally, another exception that is found in its native state, when it is found as iron meteorites. Iron in this form was worked, with difficulty, by the ancient Egyptians, who prized it greatly for its rarity (Tutankhamun was buried with some iron artefacts as well as his magnificent gold ornaments). When the technology of iron smelting from ores developed, most iron ores were small-scale local deposits, and such exploitation continued for centuries. The Victorians during the Industrial Revolution, for instance, exploited scattered iron carbonate concretions that could be found associated with coal seams, or thin iron-bearing limestones that formed in shallow Jurassic seas.

Virtually all of these iron ores have now been superseded, as a genuine monster has taken their place, which will ensure that, whatever else we might lack over coming centuries, there will globally be no shortage of iron and steel. Geological study of the Earth's oldest terrains, dating back two and three billion years ago, has unearthed giant iron ore deposits, far greater in scale and purity than anything formed subsequently. These are the *Banded Iron Formations*, so well-known as to have their own acronym, BIFs. The BIFs formed at the time when the Earth was making a long, complex transition from a planet lacking free oxygen at the surface, to the oxygenated one that supports all complex multicellular life today, and massive deposition of iron ores in the oceans was one consequence. We owe the present-day abundance of cheap iron and steel to a major planetary transition, an insight which has guided the prospecting for iron ores ever since.

The BIFs formed as layers on an ancient sea floor, which makes them geometrically relatively straightforward for the geologist, once located within rocks of appropriate age. Other metals are also precipitated from water as various oxides, sulphides, carbonates, and other chemical combinations, but typically deep underground, from water that is hot simply because of the depth within the ground, or because it has been heated by magma. These are thus *hydrothermal* deposits, which typically precipitate within complex networks of fractures to form various kinds of *mineral vein*; this represents quite a different challenge to the geologist.

The metals commonly formed in such settings include copper, zinc, lead, tin—and gold. Quite where they come from is often one of the puzzles for a geologist. Let us say that a major volcano has formed somewhere. This will bring with it magma, which will release some of the primordial water it has carried with it from the mantle, as a superheated fluid that will carry dissolved metals, that may go on to be concentrated as ores. But the volcano is also a gigantic heat engine which will heat the normal groundwater within the surrounding rocks, which in turn originated as rainwater from the sky. This re-heated *meteoric* water will also circulate around the roots of the volcano, and may dissolve metals from the surrounding rocks, to precipitate them as mineral deposits elsewhere. So—when the geologist examines a mineral vein full of metal ores—where have these ores been derived from? To answer such questions needs close study of the specific minerals, and of the order in which they formed, and of their chemistry. The isotopic compositions of such elements as oxygen and sulphur within the ores are often particularly revealing, because they may be able to show whether the components of a particular mineral were assembled from magmatic or meteoric water (Figure 30).

What causes a metal to come out of solution to precipitate as a mineral ore? This may be a change in temperature, as the

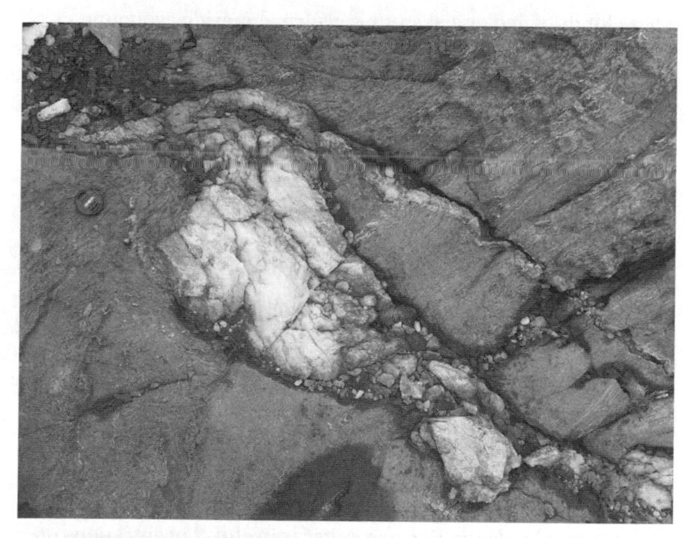

30. A mineral vein, where quartz (white) and other minerals have crystallized from subterranean hot waters circulating through rock fractures. Cap Creus, Catalonia.

mineral-rich waters cool as they ascend. Or it may be a change in pressure. It seems that some metal ores are formed when the superheated underground waters carrying the metals boil as their pressure suddenly drops, as if they were caught up in some over-stressed and unreliable plumbing system where some part of the pipework has just sprung a leak. Another potential trigger for ore formation is through the metal-carrying waters coming into contact with waters of different chemistry underground, to lead to rapid mineral precipitation. These are some of the factors that an ore geologist has to tease through, even while trying to resolve the complex geometrical patterns of the rock fracture systems in which the vein deposits are formed, and which reflect shifting underground stress patterns. The study may not stop there, as many such mineral deposits are further re-fashioned near the ground surface as *supergene* deposits, in effect the result of metal re-distribution processes associated with weathering.

These kinds of geological studies evolve alongside economic needs. Many modern technological devices today, including computers, mobile phones, and wind turbines, need *rare earth elements* such as neodymium and europium to function properly. These elements are not particularly rare in the crust, but are notoriously difficult to separate out and concentrate into forms that can be exploited. Thus, the few natural situations where rare earth element minerals are concentrated have become a focus of study, one such focus being a rare kind of igneous rock that is very carbonate-rich, and so is now a target for prospectors.

Mineral and metal exploration needs will continue to evolve further, for sure. As humans are now exploring other planets and other moons, the possibilities of prospection on such worlds will arise. This will be fascinating to follow, though it may well be that other planets, moons, and asteroids are not anywhere near as well-endowed as is the Earth as regards useful concentrations of diverse metal ores. The key to many of these ores is abundant amounts of hot water circulating underground, to strip out the metals that are present in low amounts in ordinary rock, and re-distribute them to a few places where they are present in high amounts as ores. As regards the Solar System, the Earth, where this kind of machinery is maintained by the motor of plate tectonics combined with abundant liquid water at the surface, may long remain the supreme metal factory, even as our spacecraft go on ever more distant prospecting missions.

The building trade

Each person born into this modern world arrives with an order form—attached to the forehead, perhaps—for some 500 tons of sand and gravel. The order form is admittedly a small flight of fantasy, but the physical reality of this mass of sediment is not in question, albeit as an average. One can add to that figure some several tens of tons of mudrock, limestone, crushed rock, bitumen, and other materials. These all make our share of the buildings we live and work in, the shops where we buy our goods, the roads we travel on, and the airports we take off from to go on holiday. All that material has to come from somewhere, and it lies within the realm of that part of geology that is *industrial mineralogy*.

Industrial minerals and rocks are needed in large amounts, and transport costs are high relative to the intrinsic value of the material, so much needs to be excavated relatively near to where it is used. Therefore, most places where people live in abundance must have, somewhere not too far away, many millions of tons of the bulk materials to build infrastructure with. Luckily, this is generally the case.

Sand and gravel makes up the bulk of the favourite modern building material worldwide, concrete, of which something like half a trillion tons have been produced since the mid-20th century—enough to place a kilo of the stuff on every square metre of the Earth's surface, land and sea. Sand and gravel can be arrived at, geologically, by a number of routes. The basic recipe is simple.

When rocks are eroded, they disintegrate into a mass of coarse sediment (boulders, gravel, and sand), finer material (mud), and chemicals in solution, such as carbonate and chloride ions. As this mix of sedimentary detritus is then washed by wind and water across the land, and then through river systems into the sea, it is winnowed and sorted: the heavier material tends to stay behind, while the lighter material carries on moving downstream. Certain parts of this continually active sedimentary conveyor belt concentrate the sand and/or gravel, such as river channels, wind-formed dunes, beaches, and tide-swept shallow seas (with mud being swept away, to come to rest in calmer conditions, where it can form useful accumulations that can be quarried to make bricks).

To extract these materials for building, one can simply go to modern examples of these environments and start digging. In some places, this kind of extraction takes place—a good deal of sand and gravel is dredged from modern shallow sea floors. But generally such exploitation would damage an environment that people need for other purposes, as many beaches are needed for tourism. Therefore, what is more commonly done is to look for recently fossilized examples of such environments, where the competing social pressures may be lesser. These cannot be too old, for very ancient examples are mostly lithified into hard rock, which may have other purposes (as building stone, say), but is not so convenient for concrete manufacture.

These (geologically) recently abandoned river courses, beaches, and shallow seas are part of what geologists map and analyse during the mapping of 'drift' deposits. Particularly useful forms here are river terraces, ancient floodplains perched on valley sides, and the remains of the meltwater streams that gushed out of the ice-sheets that covered wide landscapes during the Ice Ages. These are not so regular in form as the river terraces, but the constantly changing glacial channel-forms in the harsh glacial environment were often highly effective at winnowing the mud away from the sand and gravel, to produce a deposit that—long after the mammoths had disappeared from the landscape—could provide shelter and employment for the distant descendants of the mammoth-hunters.

Food and drink

The history of the human race has been a long struggle to secure not just shelter, but enough food and drink to keep alive. This has been constant, from the days of hunter-gatherers, to the early agrarian societies, to today's industrialized society. Even as human populations have grown enormously through these phases, enough food had to be found or produced. While the agriculturalists developed new technologies to achieve this, a consideration of geology was needed—and is increasingly necessary—to help maintain the boundary conditions of sufficient water and nutrients.

Some of the most famous of the geological pioneers thought heavily about the nutrient side of things, as they lived through a time when human populations began to grow rapidly. John Henslow was Charles Darwin's beloved mentor when Darwin was an undergraduate student at Cambridge University (Darwin later said of him that 'a better man never walked the earth'). Knowing that modern manure was good for crops, Henslow experimented with fossil manure—fossilized animal droppings common in some rare rock layers, together with the fossil animal bones—and found that these were effective, too, and so encouraged the local farmers to use this prehistoric resource. The Reverend William Buckland, Dean of Westminster and Oxford geologist, developed this natural resource further (he it was who introduced the term 'coprolite' for this stuff), and energetically publicized the way that the German chemist Justus Liebig improved the process by chemically processing the coprolites, which were a natural source of phosphates. Buckland grew a turnip a yard in circumference as demonstration of the effects of such super-fertilization. He had a strong personal interest in food, and was said to have tried to eat every kind of animal then known (bluebottles and moles, he said, were the least palatable).

Phosphate is still needed as an essential nutrient, and is still avidly sought in the rocks. The strata most exploited today are marine sedimentary rocks called *phosphorites*. These mostly formed in ancient seas that were highly biologically productive, often in *upwelling areas* where deep nutrient-rich waters came to the surface to stimulate intense biological productivity (a modern example occurs off the western coast of South America, where plankton abound to sustain extraordinarily productive anchovy fisheries). These are rich deposits, but only sporadically present on Earth. Morocco, by a quirk of geology, has much of the world's supplies. Just as people talk of 'peak oil' when demand outstrips supply, so there has been recent discussion of 'peak phosphate', which will be arguably even more critical to future human life (new energy sources can be developed, but there is no substitute for this particular chemical). Averting or delaying 'peak phosphate' will depend on the geologists' skill at finding more phosphorites, as well as on the effective conservation of the deposits so far identified.

Water is perhaps the ultimate necessity for life, and while Earth is a blue planet, more than half covered by oceans, the supplies of fresh water are much smaller, and ever more under pressure by a growing human population. Freshwater supplies include surface sources, in lakes and rivers—now hugely augmented by the reservoirs that have been built across nearly every major river in the world. In many parts of the world, though, much or most of the water supply comes from underground, and study of this resource forms the part of geology that is *hydrogeology*.

There are often considerable advantages to using groundwater. Its supply does not depend on annual cycles of rainfall and drought, and it is often at least partly shielded from the surface pollution that can make surface supplies unusable without expensive treatment. Locating and monitoring underground water is not a simple exercise, though, bearing some similarities to looking for oil and gas underground. Water, like these other fluids, can only be effectively extracted from rocks that are both porous and permeable.

The water-bearing rock units or *aquifers* may be layers of sand or gravel, or they may be rock which superficially seems impermeable, but which in fact stores and transmits water easily because of extensive interconnected fracture networks. The chalk that underlies London is an aquifer of this kind, and formerly contained so much water that, if a borehole was drilled into it, it became a *flowing artesian well*, with the pressurized water gushing to the surface—though soon so many boreholes were drilled to supply the city, that the water level dropped, to stop this kind of artesian flow. Another aquifer of this kind is made of thick basalt layers near the summit of the island of Tenerife. Basalt is a rock that in a museum specimen usually seems completely impermeable, but these basalts are so fractured and jointed and saturated with water that they provide an effective supply for the island—even during the tourist season.

Most aquifers, though, are more conventional bodies of porous and permeable sediment. To work out their three-dimensional shape (and so their capacity to store water) the geologist has to reconstruct them as elements of a landscape that built up and evolved over time, and link them to the processes—often dramatic—that constructed them. Some, quite hidden from view at the surface, are wonderfully useful once located. Beneath much of the plains of central Europe there are networks of spectacular 'buried valleys'—steep-sided canyons up to 400 metres deep, carved by sporadic but catastrophic torrents of meltwater unleashed from the huge ice-sheets that lay across the whole region. The lower parts of the valleys are filled with sands and gravels laid down and winnowed by the onrushing waters, and they form capacious aquifers. Best of all, above this water store, there is typically a thick bed of clay, laid down from calm waters once the canyon system had become a string of lakes, when the ice had melted. This clay shields the aquifer from view at the surface—and also from the pollution that often makes the groundwater at shallower levels unusable. If the aquifer is exploited sensitively and carefully, the waters can stay pristine for a long time.

Such pollution is one of the many hazards associated with geology. There are a large and, alas, growing number of these.

Chapter 8
Geology for society and the environment

Geology as a science encompasses the whole planet, inside and out. Small wonder, then, that geology is involved, in one way or another, in most of the resources that sustain us. By the same token, it is a significant factor in many of the hazards that threaten us, and geology is often involved in devising means of avoiding, or counteracting, or simply of living with, these hazards as sensibly as possible.

The hazards have to be taken into context. The Earth is a very smoothly running planetary machine, by any cosmic standards. It is characterized by the continuous action of plate tectonics, a unique mechanism, at least in this solar system. This process involves some very major crustal rearrangement, as whole oceans split apart, releasing white-hot magma to the surface, while slabs of lithosphere some 200 kilometres thick are forced downwards, sliding thousands of kilometres into the depths of the Earth, shouldering their way past slabs of similar thickness. Intuitively, one might expect any planet subject to such permanent and wholesale re-fashioning to be a thoroughly precarious and dangerous place. However, the machinery operates quietly and efficiently—for the most part—and has done so for billions of years, allowing our planet's surface to be continually inhabited by an array of living organisms over that time.

Nevertheless, hiccups in the smooth running do generate hazards such as earthquakes and volcanic eruptions, and these can be assessed by geological study. As well as these dangerous Earth-generated phenomena, there are yet others that we humans are answerable for. For humans are now numerous and powerful, and have become a geological force in their own right. Some of the hazards that we generate, such as chemical pollution and changes to climate, also need to be monitored, so that we can minimize the changes that result, or adapt to them as well as we can.

Earth-generated hazards: volcanoes

In the spring of 1991, swarms of small earthquakes began to be felt around the old volcanic edifice of Mount Pinatubo, on the densely populated island of Luzon in the Philippines. The volcano had lain dormant for half a millennium, but these earthquakes, monitored by a team of geologists from the Philippine Institute of Volcanology and Seismology and the United States Geological Survey, indicated that magma was rising beneath the volcano. Over the next few months the earthquakes grew stronger and more frequent, while large amounts of steam and sulphur dioxide gas were emitted. In early June, a mass of magma forced its way to the surface; it had lost its gas and, shorn of this propellant, emerged only as a dome of stiff lava. A few days later, on 7 June, a large blast took place, sending an ash column 7 kilometres into the sky. The geologists, who had been monitoring events closely and anxiously, issued a warning that a major eruption may happen within two weeks.

The timing of the statement was crucial, and agonized over by the scientists. Six million people lived around the volcano. What was then the largest US air base in the world outside the USA, Clark Air Base, was just 14 kilometres from the volcano's summit. If a call for evacuation came too early, then some of the population, tired of waiting and becoming cynical of official warnings, might return just as the eruption struck. If the call came too late—well, that scenario would not bear thinking about. The team worked hard to convince the population of the threat, and familiarized them with the evacuation zones that had been set up and with the different levels of alert, which changed from day to day. The first evacuations were called for on the day of that first eruption, in the areas closest to the volcano, and successive zones were evacuated over that week, as further earthquakes struck and more blasts took place. The call for evacuation of the air base took place just before the climactic eruption; the volcanologists had placed their calls perfectly.

On 12 June, the Philippines' Independence Day, the second largest eruption of the 20th century took place. It ejected some 10 cubic kilometres of magma and ash, forming an eruption column that rose 34 kilometres, darkening the sky. Ash and fist-sized rocks fell from this cloud over many thousands of square kilometres. To make matters worse, a typhoon, by chance, swept in from the sea as the eruption took place. Ground-hugging pyroclastic density currents poured out from the summit, filling the valleys with super-heated ash to depths of hundreds of metres. The volcanologists left their post at the last moment, driving in pitch darkness through falling ash.

The landscape and infrastructure were devastated, and 847 people died in that event, mainly as roofs laden with sodden ash collapsed (a greater number were to die in future years, whenever heavy rainfall dislodged loose ash, causing it to sweep downslope as deadly volcanic mudflows or *lahars*). Without the close monitoring of the volcano and the effective organizing of the societal response, many tens of thousands of people would have perished.

Volcanoes are a hazard that generally give warning—and also provide a useful history. Each volcano is surrounded by lava deposits and by layers of ash, which can be studied to work out when and how eruptions happened in the past, and how far their

31. A group of volcanoes with well-defined craters and eroded flanks, in Indonesia.

destructive effects were felt. This is in effect a prehistory, discernible even in volcanoes where there are no historical records of any eruption, and can help the drawing up of plans for responses to any resurgence of activity. And with each eruption, volcanologists become a little more savvy, and better prepared for the next one (Figures 31 and 32).

The analysis of ancient volcanoes, though, shows that eruptions in recorded history, including Pinatubo, were minor events compared with the super-eruptions that have taken place in the geological past—thankfully infrequently. About 75,000 years ago, eruption of the Toba volcano on Sumatra ejected somewhere between 2,000 and 3,000 cubic kilometres of ash into the air, and the resulting effects would have included a global 'volcanic winter' due to fine ash injected into the stratosphere, blocking sunlight. There were three eruptions of comparable scale in the last 2.1 million years from the Yellowstone caldera of North America (a caldera being the subsided remnants of a volcano, once the magma has been ejected). If such an eruption happened tomorrow, it would devastate the entire continent for many years. Luckily, the chances of this happening at any particular time are small.

Earth-generated hazards: earthquakes and tsunamis

Earthquakes, unlike volcanoes, typically strike without warning. Even where there are precursor earthquakes, these are usually only precursors in hindsight; at the time they occur, they are not readily distinguishable from the common small earthquakes in seismically active regions, which cause little damage. What then, can geological, or more specifically, seismological science, do to help societies at threat?

Although prediction has made limited and slow progress, the systematic gathering of seismological data on the many earthquakes that occur has identified not just the main areas of the Earth that are earthquake-prone, such as the Pacific 'Ring of Fire', but also more detailed maps of the origins, or *epicentres*, of earthquakes, that may be linked with the detailed geological mapping of the network (often complex!) of the major fracture lines or geological faults in the Earth's crust (Figure 33). Such mapping is not easy, especially where the bedrock is poorly exposed in deeply soil-covered or vegetated regions. Some faults have been revealed, even in recent years, only when a sudden, unexpected earthquake takes place.

Nevertheless, once such major fault systems have been reasonably well mapped out, and combined with detailed local records of earthquakes, *quiet segments* may be recognized along them: zones that have not experienced any earthquakes in recent years, and so may be assumed to be building up stress that may indicate that they are more likely than other fault segments to rupture, triggering a major earthquake. The slow displacement of crust that indicates such stress build-up may also be tracked by the astonishingly precise (centimetre-scale) measurements of crustal

32. The Stromboli volcano in Sicily erupting, photographed at night.

displacement now possible by techniques such as *laser interferometry*. Nevertheless, these studies still cannot predict precisely (or even approximately) when an earthquake will take place, not least because a significant amount of the movement of rock along fault planes takes place not by sudden shock-generating movement but by slow *aseismic fault creep*. The studies might, though (rather tentatively), indicate districts where there may be greater or lesser chances of earthquakes along such fault zones.

However, useful geological information can still be provided that can help in the design of built infrastructure to make it more earthquake-proof. For instance, areas that are underlain by soft unconsolidated deposits can be discriminated from those where hard bedrock comes to the surface; the former generally need more engineering protection, as such soft deposits can amplify the shaking effects of an earthquake. And, in mountainous areas, areas can be mapped out that are likely to be in the path of tumbling boulders, dislodged from rocky crags above. In earthquake-prone areas, though, the best protection still comes from appropriate building design and construction. Buildings that

33. A tectonic fault, Tabernas, Spain. Where the prominent dark sandstone layer (on which the students are standing) disappears, the strata have been cut across and displaced along a fault plane (the surface trace of which runs across the middle of the photograph).

have lightness, strength, and flexibility built into them can survive far better than those that are heavy and brittle. This is the major reason why an earthquake in a rich country can cause relatively few casualties, while a similar earthquake in a poor country can kill many thousands of people.

Earthquakes that take place near the coastline may generate *tsunamis*, as the sudden lurching of a section of sea floor generates a train of waves large enough to cross an ocean and devastate coastlines thousands of kilometres away. The key for survival here is preparation. The wave travels across the open ocean at up to 1,000 kilometres an hour—and so extremely quickly, but giving at least the more distant locations time enough for warning. When an earthquake strikes, seismological stations can, in minutes, pinpoint its location and strength, assess its tsunami-raising potential, and decide whether to send an alarm. In the Pacific region, there has been a tsunami-warning system in place for several decades, and there are education programmes to advise coastal populations on what to do when an alarm sounds.

In the Indian Ocean, tsunamis are less frequent. Prior to 26 December, Boxing Day, in 2004, the last major one had been triggered by the eruption of Krakatoa in 1883. That one killed some 36,000 people, but a century is a long time in the memory of modern human cultures. So there was no such alarm system, and no public education systems in place, when a huge earthquake—the third largest ever recorded on seismographs—struck off Sumatra, as a tectonic plate boundary ruptured over a length of some 1,600 kilometres. The resulting tsunami devastated coastlines successively from Indonesia, to Sri Lanka and India, to the east coast of Africa, killing some quarter of a million people in all. Many people on the coast did not associate the drawdown of the sea, a few minutes before the main wave struck, with a tsunami, and so did not use that time to run for higher ground—some even wandered down onto the suddenly-exposed sea floor, curious at what had happened. (Some indigenous peoples fared better, recognizing and using this natural warning from folk legends of tsunamis from many generations ago.)

Since then, a tsunami warning system has been put into place for the Indian as well as the Pacific Ocean. But it is not only earthquakes and major volcanic eruptions that can generate tsunamis. In peat bogs along the north-east coast of Scotland, there is a layer of sand, pebbles, and shells that dates back 8,150 years. It represents the traces of a tsunami triggered by an enormous undersea landslip scar off the coast of Norway, rendered visible today by sonar imaging. Some 8,150 years ago, a slab of shallow sea floor, nearly 300 kilometres in length and involving over 3,000 cubic kilometres of rock and sediment, suddenly slid into deep water. The resultant tsunami from what is now called the Storegga Slide would have devastated the communities of early hunters—not least because it struck, for them, at a fatally wrong time of year. Fossilized plants in the tsunami debris show that this catastrophe took place in late autumn, when the hunting communities would have come down from the hills, unsuspectingly, to over-winter by the coast.

Geologists studying volcanic islands such as Hawaii and Tenerife now recognize similarly huge landslip scars on the sides of these islands, and the debris deposits offshore are often bigger in area than the islands themselves. These islands probably also generated huge tsunamis when parts of them suddenly collapsed, thousands of years ago. Useful assessment of the risk of future events depends on gaining a better understanding of these ancient catastrophes, by forensic analysis of the geological deposits they leave behind.

Human-made geological hazards

There are many other geological hazards as well as those from volcanoes, earthquakes, and tsunamis. Deep holes can open in the ground, after limestone or salt strata have been dissolved underground by groundwater. Floods can sweep through towns and cities. Snow or rock can avalanche from steep mountainsides. Groundwater may contain natural contaminants such as arsenic. But there are also a gathering number of geological hazards caused or threatened by our own activities, and these are increasingly becoming a major focus of attention for geologists. The scale of some of these hazards promises—or threatens—to become of planetary scale, and to imperil our own future as a species.

The carbon release

Analysis of the geological record shows that the world's carbon cycle is finely balanced, and fundamentally underpins conditions of climate and life on our planet. The ice strata drilled from the centre of Antarctica extend back 800,000 years. Over that time, the tiny bubbles of fossil air in the ice show that the amount of carbon dioxide in the air has oscillated regularly between about 180 and 280 parts per million (ppm), and this pattern matches well-nigh perfectly with times of glacial conditions (low CO_2) and warm interglacial conditions (high CO_2)—unsurprisingly, because carbon dioxide has been physically shown to be a greenhouse (heat-trapping) gas. The same record, examined in more detail, shows the last of these natural oscillations as a more or less steady rise in CO_2 concentration taking some 6,000 years, as global temperatures increased into the current interglacial phase. Then, within this phase, for nearly all of the last ten millennia or so, both CO_2 levels and global temperature stayed generally stable—a key factor in allowing human civilization to develop. Looking more closely, there was a very gentle, almost imperceptible, increase in atmospheric carbon dioxide levels from about 7,000 years ago; this has been ascribed, controversially, to the farming activities of early human communities, and may have been sufficient to prevent the Earth from sliding back into a glacial state.

The rise from this trajectory, at first from the late 18th century, then more markedly from the mid-20th century, is now precipitous—becoming over a hundred times faster than the rise in carbon dioxide levels at the end of the last Ice Age. The level it has now reached—over 400 ppm—is far greater than at any time over the past 800,000 years, and such carbon dioxide levels were probably last seen some three million years ago, when the world was warmer and sea levels were some 20 metres higher, as icecaps were smaller and stored less ocean water than they do today. What is happening is in effect a reversal, mostly within a couple of centuries, of hundreds of millions of years of burial of carbon as coal, oil, and gas. One set of geological skills located those ancient carbon stores and enabled their exploitation in a geological instant—and another set of geological skills is now allowing the magnitude of this action to be assessed at a planetary scale (Figure 34).

Other geological contexts allow the unintended consequences of this huge carbon release to be assessed. One consequence is

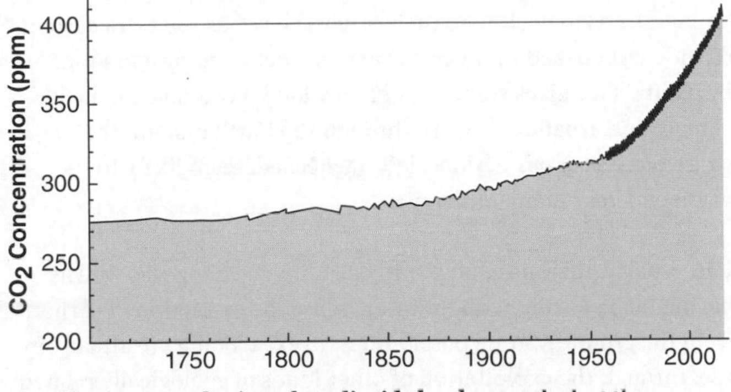

34. Rise in atmospheric carbon dioxide levels over the last three centuries.

already clear just from present-day measurements, with no recourse to geology: global temperatures have risen by a little over 1 degree Celsius so far in the last century, while sea levels have risen by about 20 centimetres in that time; both are on an upward trajectory. How far can these go?

The geological context can help inform longer-term consequences, over centuries and millennia. Within the Ice Ages, the last interglacial phase similar to our own was 125,000 years ago. With carbon dioxide levels at 280 ppm, temperatures in this phase were similar to those of today, while sea levels reached up to 5 metres higher than the present level. This suggests that, even if carbon dioxide levels stabilized at 400 ppm (as I write, they are rising at about 2 ppm each year), both temperature and sea level may keep rising for some time.

There is another, considerably older, geological example, which may act as a partial guide to the future. Fifty-five million years ago, at what is now seen as the boundary between the Paleocene and Eocene epochs, the world rapidly warmed by some 5 degrees as large stores of carbon were released from the ground into the atmosphere and oceans. It took about 100,000 years for temperatures to decline to their original level, as the extra carbon dioxide was soaked up by reaction with rocks and by extra take-up by plants. This gives some idea of how long it can take for such a climate perturbation to work through the Earth system; the consequences of our actions, left unchecked, seem likely to persist for very many millennia.

Can geology, therefore, help mitigate climate change, as well as having helped bring it about by enabling the extraction of carbon from the ground? Some possibilities do exist. Some are already in use, through the exploitation of other kinds of geologically related energy. Hydroelectricity is already a mature technology, with most major rivers in the world having dams somewhere along their courses, generating electricity; construction of these dams needs both considerable *engineering geology* expertise to site and construct the dams safely, and the location of geological resources (to provide large amounts of concrete, for example) to build them. It is possible to exploit hot rocks underground, too, through *geothermal energy*, by pumping cold water underground and then extracting it once it has been warmed by the hot rocks, especially in volcanically active areas like Iceland. Neither of these forms of energy generation are problem-free: dams trap large volumes of nutrient-rich sediment and the drowned vegetation releases methane, a greenhouse gas; and the water circulated in geothermal energy soon becomes corrosive to the machinery used in energy generation. As the methods develop, ways may be found to resolve such problems.

Another approach is by *carbon sequestration*, to put carbon back into the ground. One form of carbon sequestration involves using the skills and knowledge developed in fossil fuel exploration, only in reverse; this involves injecting carbon dioxide under pressure into exhausted oil and gas reservoirs underground. Another promising avenue is to accelerate the process of rock weathering to soak up carbon dioxide, using the large volumes of crushed rock waste generated by the mining industry. Yet another is to extract magnesium-rich igneous rocks, and react them with carbon dioxide on an industrial scale. These, and other such techniques, are currently in small-scale, exploratory form. Time will tell if they develop sufficiently to help stabilize climate.

The waste stream

In part due to its gargantuan energy use, modern human society has developed the means to manufacture large volumes of material objects of all sizes and types, from city skyscrapers to cars and aeroplanes, to household goods and their packaging. Most of these objects have a limited lifespan, and at the end of their use they are recycled, thrown away, or incinerated. Recycling, though, is in many cases really 'downcycling'—re-fashioning the substance into something of lower quality that, once finished with, cannot be recycled again. And incineration generates a good deal of ash, which must be disposed of, often in landfill. Most things we make, therefore, are eventually thrown away. So while the aim of the manufacturing industry is to get closer to a 'circular economy', with continual re-use of materials, in reality much of what we now use will be thrown away, to find its way eventually into landfill sites. And, here again, geology comes into play. For those landfill sites need to be designed to cause minimum harm to their surroundings, and when waste does leak out, its effects must be studied to gauge the level of damage to the environment.

There is a symmetry of a kind between resource extraction and waste disposal. The large quarries excavated for such resources as sand and gravel and brick clay can have a second economic life (often more profitable than the original one) as a landfill site, once mineral extraction stops. However, the geology of the site must be considered to find out what kind of waste can be dumped into it, and what precautions need to be taken. For instance, sand and gravel quarries are excavated in very permeable strata. If the material dumped in them includes toxic chemicals, these can all too easily leach out and pollute groundwater supplies and soils over a wide area. Hence, to dispose of such substances, it is better to site them in old clay pits, where the geological strata are impermeable, and toxins can be contained. Increasingly, such sites are also engineered, being lined with tough impermeable plastic sheets, to contain all of the waste material within. This can set up its own problems as, in these enclosed conditions, organic waste material does not decay naturally, but slowly ferments to release methane, a potent greenhouse gas. Resolving this problem needs further engineering work, in collecting this gas and burning it; if this is done properly, the methane problem can be turned into an energy resource.

The rapid development and innovation of technology means that the nature of waste material—and of the problems it raises—is constantly evolving too. Before the mid-20th century, plastics were virtually unknown. Since then their use has grown enormously. Currently each year something like 300 million tons—a mass about equivalent to that of the world's human population—is manufactured. Most of the plastics ever produced, estimated to be more than 8 billion tons (easily more than enough to wrap the whole world in a layer of clingfilm, or plastic wrap), are still present in the world. As they are tough, light, and highly resistant to decay—all the properties so useful to humans—plastics carelessly disposed of have now dispersed across all inhabited landscapes, and through all of the oceans. These abundant new particles are hazardous, choking the wildlife that eats them and absorbing toxins on their surfaces. Even remote sea floors and far-distant beaches now contain plastics, all too often abundantly, as a new component of strata forming today. Geologists are among the scientists now studying this new environmental problem, helping to track the pathways and

35. Plastic litter. This image represents an approximate global average (land and sea) of the plastic made and thrown away.

ultimate fate of plastics, and how this new phenomenon will evolve in the future (Figure 35).

Not all environment contaminants are as visible as plastics. Another post-Second World War development has been a suite of long-lived organic chemicals such as pesticides, which again have become dispersed across much of the world. There are also dispersed metal particles and compounds from mining operations, smelting plants, and factories—and radioactive particles from nuclear power plants. A modern geochemist is as likely to be tracking these kinds of signals in river, lake, and estuary sediments, as to be analysing the chemical compositions of ancient rocks and minerals.

As these pollutants change and evolve, and build up in the environment, there will be an ever-greater need to study them, to help devise the best means for society to deal with them. New circumstances, and new combinations of circumstances, will certainly arise. For example, the interaction of climate change—as sea level rises and patterns of rainfall and erosion change—with pollutant stores held in shallow stores such as landfill sites will need close attention. There will be a lot of work to do.

Chapter 9
A very brief history of the Earth

Our planet is ancient—at 4.6 billion years, it is almost exactly one-third of the age of the cosmos. Over that time, it has changed enormously—indeed, it has rather been a succession of different planets than a single planet. We humans would not be able to survive on most of those planets.

The first of these planets is now unknowable in any practical sense. All we know is that it would have been so different that it would not have had any similarity to the planet that it subsequently transformed into. It has, therefore, been given a different name by some geologists: *Tellus* is the planet that first formed in Earth's orbit, as our Solar System arose from the gas and dust that circled the infant Sun. For some tens of millions of years, Tellus evolved in some now unknowable way, in an eon that those same geologists called the Chaotian. Then, in one moment, Tellus vanished as another planet, that has been called Theia, collided with it. Both Tellus and Theia were utterly destroyed, and the smashed material reorganized itself as the Earth and the Moon. This narrative is the only sensible way to explain the Earth–Moon system as it currently is, while the close chemical similarity of the Earth and the Moon suggests that the orbit of Theia was very similar to that of Tellus—and so this was an accident waiting to happen.

The Earth in its first half-billion years, though, is almost as mysterious as was Tellus. This time is called the Hadean.

The Hadean Eon

The Earth has re-invented itself so completely since the Hadean that almost nothing remains from that time to be analysed. The main exception comprises some of the most famous crystals in the world—within the community of geologists, at least. They do not look special at all—a nondescript grey, and less than 1 millimetre in size: they are crystals of zircon. When radiometrically dated, these particular zircons (less than a thimbleful have been found) were found to be truly ancient, as some of them are over four billion years old, the oldest so far found being dated at 4.4 billion years, and so from very near the beginning of Earth. They come from the Jack Hills in Western Australia, and are found within sandstones that are not quite so ancient, having formed just three billion years ago, but including detritus from yet more ancient terrains, of which the rare zircons are now the only trace.

The Jack Hills zircons have been intensely studied, to provide clues about the world of the Hadean. Even though they crystallized within magma several kilometres below ground, they contain faint clues to conditions at the surface. The proportions of isotopes of oxygen atoms in the crystals have been interpreted as suggesting the presence of water on the Earth's surface at that time. Even more remarkably, one zircon, 4.1 billion years old, contains a scrap of carbon as graphite that, when analysed, showed structures and chemistry (again, as the proportions of isotopes among the carbon atoms) that suggested—*very* tentatively—that it had been processed by some kind of life before it found its way deep underground.

Any kind of life present in the Hadean would have had a most precarious existence on this young planet—not least as the young

Earth was pelted by meteorites in the debris-rich early Solar System. A particularly heavy influx, the Late Heavy Bombardment, is thought to have taken place about four billion years ago, perhaps triggered by a rearrangement of the orbits of the large planets. Almost all traces of these impacts have now disappeared on Earth, through tectonism and erosion, but the impact-scarred face of the Moon gives an idea of the scale of these tumultuous early events.

From about 3.8 billion years ago, rocks began to be preserved. With evidence from these, a clearer picture can be built of the early Earth. This is the world of the Archean Eon.

The Archean Eon

The oldest Archean rocks include some that formed at the surface, as sedimentary rocks. Many have been strongly metamorphosed since, during their long sojourn deep in the crust; nevertheless, they tell something of what surface conditions were like in those times. There are strata that were laid down from running water, in rivers and in some kind of deep sea. There was life, wholly microbial but abundant enough as microbial mats to construct, by trapping sediment, finely layered rock structures called *stromatolites* on the sea floor. But it was not a world for humans. The sedimentary grains transported by rivers across the land surface included minerals such as pyrite (iron sulphide) and uraninite (uranium oxide). Today, these do not survive for long at the surface but soon rust away—therefore, the Archean atmosphere lacked oxygen. So, what did that atmosphere contain?

One clue comes from a puzzling absence among the early Archean strata. There is no sign that there was any significant ice at the surface. The signs of such ice are not subtle: glaciers and ice-sheets grind rock and soil into a mixture of clay, sand, and boulders that show characteristic scratches and grooves on their surfaces from being dragged along the ground by the ice. This very characteristic kind of deposit, termed a glacial till, is common across parts of the world that have been affected by the recent Ice Ages, but no fossilized examples have been found in the first billion years or so of the Archean. The climate, thus, was consistently warm (and some geologists have suspected that at times it might have been as warm as a hot cup of tea). This is puzzling because the Sun would have given out some 20 per cent less light and heat than it does now. The best explanation of this apparent contradiction is that the Earth then had more greenhouse gases such as carbon dioxide and methane in its atmosphere, to stay warm in the faint rays of the young Sun.

The Archean is also suspected to have seen a much more deep-seated planetary change: the beginning of modern-style plate tectonics. The evidence for this is patchy and tantalizing, but the hints include the appearance *within* diamonds—which form at very high pressures, hundreds of kilometres below the Earth's surface—of tiny specks of mineral of the kind that form in subduction zones. This appearance took place a little more than three billion years ago. Before that, what kind of tectonics did the Earth have? Some geologists have suggested that there was a faster-moving, shallower version of plate tectonics on a hotter Earth, or that enormous meteorite impacts might have triggered short-lived phases of plate tectonics in some kind of stop-start fashion. There has also been a suggestion that our planet may then have been made of a single plate, through which magma ascended in vertical 'heat-pipes'—a little like the mechanism by which volcanism today takes place on Io, one of the moons of Jupiter. There is still much study needed, to show how our planet functioned in its earliest billion years.

Towards the end of the Archean, the Earth changed into a different state—one that we would begin to find a little more conducive to life, though not comfortably so for humans, by any means. Oxygen came into the atmosphere, to create the world of the Proterozoic Eon.

Proterozoic Eon

Some time around 2.5 billion years ago, some microbes evolved a means of harnessing sunlight to combine carbon dioxide and water to make carbohydrates, and as a by-product they released oxygen in photosynthesis. Once the oxygen began to seep into the atmosphere, the landscape changed. This chemically highly reactive gas (which was as deadly to most of the microbial population then as chlorine gas is to us now) began to react with rock minerals in what is now called the *Great Oxygenation Event*. The colours of the land surface would have turned from greys and greens to the browns and reds of rust, and minerals like pyrite and uraninite that had been protected in the Archean atmosphere now swiftly altered to oxides and hydroxides.

In the sea, there was a proliferation of Banded Iron Formations—the characteristic layered strata composed of alternating fine layers of iron oxide and silica, which we exploit to make iron and steel. Reflecting the transition from a chemically anoxic ocean (which can hold very large amounts of dissolved iron) to one in which oxygen is present (iron is almost insoluble in oxygenated water), these rocks represent the long process of purging of iron from the oceans. But some Banded Iron Formations range back to more than three billion years ago, well into the Archean (Figure 36). Was this because small amounts of oxygen were simply absorbed in the sea, without leaking into the atmosphere? Perhaps—but a more likely explanation is that microbes had evolved an early form of photosynthesis, which did not release free oxygen, and this early mechanism was involved in precipitating iron from the oceans.

The appearance of free oxygen roughly coincides with another change. A little more than two and a half billion years ago, the first signs of extensive glaciation appear in the geological record, in the form of widespread deposits of fossil glacial till. What had

36. A characteristic Archean rock: a tectonically folded Banded Iron Formation from Minnesota, USA ~2.69 billion years old.

caused this cooling of the Earth—even as the Sun was slowly becoming hotter? One plausible mechanism is that the oxygen in the atmosphere soon reacted with, and removed, large amounts of methane that had been present. Removal of this potent greenhouse gas may have been the key factor triggering the growth of ice.

Curiously, after this marked cooling, the Earth seems to have been warm for most of the long time expanse of the Proterozoic. It has been termed the 'boring billion' by geologists. Life was still present, but remained for the most part as microbes, with no marked evolution into more complex forms. Oxygen was present, though likely in much lower amounts than at present, and this is thought to have been a factor in another major transition of the oceans, from being anoxic to being mostly *sulphidic*, rather than immediately becoming fully oxygenated. Here, sulphur minerals on land were oxidized, and washed into the sea as abundant sulphates. In the deep, oxygen-poor ocean, these sulphates would have been converted by microbes into sulphide that combined with iron to form tiny crystals of pyrite, which slowly fell as a kind of fine pyrite rain into the sediments on the sea floor. Pyrite as it crystallizes takes in other elements, such as molybdenum and zinc, which are crucial trace elements for life. This nutrient sequestration has been proposed as one reason why life evolved so slowly during the boring billion: it was being starved.

Towards the end of the Proterozoic, the Earth became anything but boring. It plunged into the most extreme glaciation in all of Earth history, the time of 'Snowball Earth'. Ice covered all the main continents, reaching to the Equator and large parts of the sea. Debate has raged as to whether it was a 'hard snowball' with a carapace of ice covering everything on the planet, or a 'slushball' with significant patches of open water. These times, more technically called the Cryogenian Period, lasted over a hundred million years with two major phases of glaciation, so perhaps both forms of snowball might have been present at different times. In either case, such great expanses of highly reflective ice would have reflected most of the Sun's light and heat, and at first there was perplexity as to how the Earth could have escaped from this state (as it evidently did). The escape was very likely made via a build-up of carbon dioxide, which would have carried on being released from volcanic eruptions punching up through the ice. As levels of this greenhouse gas built up, its effect would eventually have overwhelmed the reflective effect of the ice, and the glaciation would suddenly have collapsed.

The last major glacial collapse took place about 635 million years ago, ushering in the last part of the Proterozoic Eon, known as the Ediacaran Period. This marked the beginning of fundamental change to the biosphere, though perhaps it may have been more a precursor than a beginning. In strata of this age there are curious fossils, of the first large multicellular organisms. Called the 'Ediacaran Biota', they include disc- and frond-shaped forms

37. Examples of the enigmatic Ediacaran fossils, photographed at Mistaken Point Ecological Reserve, Newfoundland; scale bar is 5 centimetres.

(Figure 37). Despite intense research, they remain a mystery—they had little or no power of movement and no obvious mouth or gut, so it is unclear how they obtained nutrition. They may have been completely unrelated to the familiar multicellular animals we know, and when these animals arrived, which included many mobile and predatory forms, the Ediacaran Biota vanished.

Phanerozoic Eon: the Palaeozoic Era

The beginning of the Phanerozoic Eon, the eon we still live in, was marked by a phenomenon that greatly puzzled Charles Darwin. After the rocks of the Precambrian, which to him seemed unfossiliferous (the Ediacaran Biota had not been found in his time, and the microbial stromatolites were not widely regarded as of biological origin), there abruptly came rocks with a plethora of obvious fossils: carapaces of trilobites (Figure 38), shells of

38. A trilobite, an example of one of the key fossil groups of the Palaeozoic Era.

brachiopods ('lampshells'), and molluscs, corals, sponges, and other organisms. This has been called the 'Cambrian explosion' (the Cambrian, beginning 542 million years ago, is the first period of the Palaeozoic Era, which is in turn the first era of the Phanerozoic Eon). We now know that this 'abrupt' change unfolded as distinct phases of evolution over some thirty million years, but this is still rapid when compared with the slow biological change of the Archean and Proterozoic. The trigger for the 'explosion' is unknown, though factors such as a further increase in atmospheric oxygen levels have been mooted. Once started, an evolutionary 'arms race' began that has continued to the present day.

The fossil record of the Cambrian shows that complex multicellular life had a bumpy beginning, with phases of increased biological diversification interspersed with crashes seen as moderate-scale mass extinction events in the fossil record. The succeeding Ordovician Period, beginning 485 million years ago, saw, over the next forty million years, a more or less steady rise in biological richness, called the 'Great Ordovician Biodiversity Event', in which the seas increasingly teemed with a variety of life. The Ordovician was terminated by a sharp mass extinction event, but in the succeeding Silurian Period, beginning about 445 million years ago, marine life soon (over some five million years, that is) recovered its lost diversity.

The Silurian Period also shows the beginnings of a serious move of life onto land—a terrain that had remained mostly barren for some hundred million years following the 'Cambrian Explosion' at sea. By the end of the Silurian, some simple plants (little more than slender stems a few centimetres high) were colonizing lower, wetter ground, with a few invertebrates such as tiny millipedes advancing across the land with them, while some early armoured fish made the transition from the shoreline into lakes and rivers.

Through the following Devonian Period, this terrestrial ecosystem developed and by the Carboniferous Period that followed, some 330 million years ago, enormous forests of (to our eyes) strange trees, including giant ferns and horsetails, but with no flowering plants, were spreading across the land, inhabited by amphibians and by such invertebrates as 2-metre-long millipedes, and dragonflies with 60 centimetre wingspans. By a quirk of geology, this single period (and hence its name) has provided our species with a good deal of the coal that it uses. This is because, as large plants first appeared, they colonized an enormous tract of subsiding swampland that stretched from what is now North America through Europe to Asia. The swampland supported and then buried countless generations of these early forests.

The buried forests represent enormous amounts of carbon, taken from the air and put underground as coal. As atmospheric carbon dioxide levels dropped by this process, the Earth cooled and went into a prolonged glaciation. A large icecap covered the southern part of the continent of Gondwana, which combined continental masses that we now know as South America, Africa, Australia, and India, these all then being in the southern hemisphere, with part in the south polar regions. As the ice waxed and waned, global sea levels rose and fell, to periodically drown and re-expose the coal forests that were growing in Equatorial regions.

The Equatorial swamplands largely disappeared as the late Carboniferous gave way to the Permian Period. This is the time when—by another quirk of plate tectonics—all of the world's major landmasses were assembled into one supercontinent, called Pangaea. The interior of Pangaea, being far from the sea, became intensely arid, and this is one factor that contributed to the swamp forests' demise. The resulting strata, which cover much of Europe, include fossilized desert dunes and salt deposits left by dried-up inland seas. These strata contain few fossils, so they do not clearly show the catastrophe that brought about the end of both the Permian Period, and of the entire Palaeozoic Era.

The Mesozoic Era

In places where strata preserve a marine history, such as in South China, evidence can be seen of the greatest mass extinction event so far in the Phanerozoic Eon. Some 95 per cent of species abruptly became extinct 250 million years ago, including entire groups such as the trilobites. What happened?

Physical and chemical clues in the strata suggest that there was an 'anoxic event' in the oceans, when much of the sea became oxygen-starved. Anoxic events were common during the Palaeozoic Era (with a few of them being associated with greater or lesser extinction events), but this one was unusually strong. There is chemical evidence too that global warming also took place, with a sharp rise in carbon dioxide levels, and of global temperatures by something like 8 degrees Celsius. The extinction event coincides exactly in time with an enormous outburst of volcanism in what is now Siberia, where about 3 million cubic kilometres of basaltic lavas poured across the landscape in just a few million years. This outburst must have led to large emissions of carbon dioxide, sulphur dioxide, fluorine, and other noxious chemicals, and is widely seen as being the cause of the mass extinction event. Its own cause has been ascribed to an ascending mantle plume impacting on the base of the crust in this region, to trigger this outpouring of lava.

Recovery from this mass extinction event, to re-establish biological diversity levels, took more than ten million years into the Triassic Period. The world that became established then included a variety of reptiles on land, including the 'mammal-like reptiles' among which were our own ancestors. The Triassic world was terminated by another mass extinction event, 200 million years ago, again likely caused by an extraordinary burst of volcanism, this time as Pangaea was beginning to break apart, in the initiation of the North Atlantic Ocean.

This mass extinction, by removing competition, was a key factor in establishing the dominance of the dinosaurs on land as the Jurassic Period began, and of marine reptiles such as the ichthyosaurs and plesiosaurs in seas in which fish abounded, as did molluscs such as the spiral-shelled ammonites and squid-like belemnites. These general conditions extended (with no great changes at the boundary) into the Cretaceous Period, when the dinosaurs maintained their dominance on land and grew yet more monstrously large, while flowering plants began to be a significant part of the landscape. The Cretaceous Period was a time of global warmth, when there were few or no icecaps, and when sea levels were 100 metres or more higher than they are today. This led to marine, and even deep water, conditions extending across much of the continental areas, with deposition of the characteristic chalk strata across much of the world. This distinctive rock type represents a major evolutionary innovation,

39. Chalk cliffs in east Sussex, UK. These represent the globally high sea levels of the Cretaceous Period, when the sea flooded across much of the continental areas, and covered them with calcareous oozes made of the skeletons of countless coccoliths (microscopic planktonic algae). The faint stripes seen represent climate cycles, each tens of thousands of years long.

the appearance of ocean-going microplankton (both plant and animal) that possess skeletons of calcium carbonate. It is these minute skeletons that piled up on the world's sea floors, year after year, to form thick chalk deposits (Figure 39).

The Cenozoic Era

The abrupt end of both the Cretaceous Period and the Mesozoic Era and the mass extinction event that killed off the dinosaurs and much else, was coincident with a major asteroid strike on Mexico's Yucatan Peninsula, sixty-six million years ago. But decline in some animal groups was evident for a few million years before this climactic event, and may have been linked with another prodigious outburst of basaltic lavas, this time on India's Deccan Plateau.

The extinction of the dinosaurs on land allowed, finally, the evolutionary development of the mammals, which had persisted through the Jurassic and Cretaceous periods as small animals, never much larger than a dog or a cat. These now began the evolutionary pathways that would lead to such animals as the elephant and mammoth on land and the blue whale in the sea—the largest animal that has ever (as far as we know) existed—and to the humans. It also saw further spread of the flowering plants, including the development of grasses and grasslands. The world was becoming modern.

The end-Cretaceous meteorite impact collapsed the global ecosystem (again, it took several million years to rebuild biological diversity levels, albeit with completely different components to the Cretaceous ones). But it did not permanently perturb global climate, except perhaps for something like a brief and temporary 'nuclear winter'. Climate continued to be warm throughout the early part of the Cenozoic Era, with lush forests in both the Arctic and Antarctic regions.

Climate changed abruptly (that is, over something like 200,000 years) thirty-four million years ago, at the beginning of the Oligocene Epoch. In this transition, an ice-sheet grew over most of Antarctica, essentially converting it into the state that we now know. The reasons are still being debated, but this transition seems coincident with a sharp drop in atmospheric carbon dioxide levels, from something like 800 ppm to something more like 400 ppm. As to what might have caused this drop, or otherwise been involved in the transition, a number of ideas have been put forward. Around this time the Himalayas were rising, after India, in drifting northwards, had ploughed into Asia, and the enormous mass of rock lifted to high altitudes might have, through chemical reactions associated with rock weathering, taken large amounts of carbon dioxide out of the air. About this time, too, the free passage of ocean currents around the Equator was being hindered as Africa was drifting towards Europe, and beginning to turn a wide ocean (called the Tethys Ocean) into a smaller, restricted one (the Mediterranean); conversely, an open seaway was opening up around Antarctica, as Australia and South America drifted away. Whatever the exact combination of circumstances, a greenhouse world now changed into an icehouse world.

The next phase of the late Cenozoic icehouse took place about two and half million years ago, at the beginning of the Quaternary Period of the Cenozoic Era, when ice grew in the northern hemisphere, over North America, Greenland, and Scandinavia, to join the long-standing Antarctic ice-sheet. The immediate cause might have been a re-orientation of ocean currents and weather patterns in the North Pacific, which led to a kind of 'snowgun' mechanism increasing snowfall on the North American continent, in large enough amounts for a large ice-sheet to start growing there. As to what led to these new conditions, one proposed link is with the joining of the Americas by the formation of the Panama Isthmus, which had happened a little earlier. As well as allowing the 'Great American Interchange', when animals and plants from the north migrated south and vice versa, this new barrier would have affected the course of ocean currents.

The Quaternary Period is marked by a number of changes around the world, as well as being the recent 'Ice Ages' of common understanding when, during cold glacial phases, ice advanced across much of northern and central Europe and north America (Figure 40). In parts of Africa, there was drying, with forest being replaced by savannah. This was likely one factor in the evolution and radiation of a formerly obscure group of primates, from which evolved a number of hominin species, including our own genus, *Homo*.

Our own species, *Homo sapiens*, arose some 300,000 years ago in Africa. For most of that time, it was as unremarkable and obscure as its sundry kin species, although it (and other related species) had learnt to make simple tools and control fire. Then,

40. A gigantic raft of chalk rock embedded within glacial deposits, that was gouged from the bedrock by a moving ice sheet and dragged into its present position. Norfolk coast, UK.

50,000 years ago, some kind of change took place that led to 'culturally modern' humans, who developed such things as symbolic cave art and more efficient communication and social cooperation. Becoming more formidable as hunters, they were likely the main factor in a wave of extinctions of large land animals, approximately halving the number of these by 10,000 years ago, near the beginning of the current warm phase of climate of the Holocene Epoch. They were audacious and skilled travellers too, reaching many Pacific islands and Australia.

About this time, farming began, and this innovation allowed the numbers of humans to rise yet higher, and to develop a number of civilizations and empires, with growing urban centres, that helped extend their reach across the planet. For thousands of years, human history unfolded, with civilizations such as those of Mesopotamia, of ancient Egypt, Greece, and Rome, of China, of the Byzantine Empire and Renaissance Europe. Through all of this cultural evolution, though, basic Earth System parameters such as the carbon, nitrogen, and phosphorus cycles stayed approximately the same.

Then, beginning in the late 18th century, the Industrial Revolution began, a phase of human population growth, technological development, and energy use that began to shift these Earth System parameters, a process that took a sharp upturn with the mid-20th century's 'Great Acceleration' of population growth, industrialization, and globalization. From that time, these Earth System parameters have been sharply perturbed, with approximate doubling of the amount of reactive nitrogen and phosphorus at the Earth's surface through intensified agriculture, and sharp increase in atmospheric carbon dioxide and methane levels, which has begun to drive rises in global temperature and sea level. At the same time, the biosphere is being markedly modified by further extinctions, worldwide species invasions, and agricultural breeding programmes, while technological evolution is accelerating markedly and artificial intelligence now in some (still limited) ways surpasses human intelligence. This is the combination of factors that has led some to call the present time the Anthropocene, a still informal term introduced by the Nobel Prize-winning atmospheric chemist, Paul Crutzen, in 2000.

Whatever the future has in store, it looks as though these changes are bringing the Earth to the cusp of its next major geological transition. Our planet seems set to enter a major new phase of its history. Determining the scale and significance of this planetary change, and of its place within the context of Earth's almost unimaginably long and varied history (and in the context, too, of how planets and moons beyond Earth have evolved) is just one of the tasks where the science of geology takes centre stage. Geologists will continue to fathom the Earth's deep past and explore how our planet has evolved to its present state. And while the present has often been seen as key to understanding the past, knowledge of the Earth's deep geological past may, too, become one of the surer guides for steering a course into the future.

Further reading

Cadbury, Deborah. 2010. *The Dinosaur Hunters*. Fourth Estate. Splendid portrayals of the fossil hunters of Victorian times.

Fortey, Richard. 2005. *Earth: An Intimate History*. Harper. Excellent on the various phenomena—volcanoes, earthquakes—that are by-products of the working of Earth's plate tectonics engine.

Hazen, Robert. 2012. *The Story of Earth*. Penguin. A fine account of our planet seen through the eyes of this distinguished and imaginative mineralogist.

Kunzig, Robert. 2000. *Mapping the Deep*. Sort of Books. Very good on the geology—of ocean floors—that is usually hidden from us.

Nield, Ted. 2007. *Supercontinent: Ten Billion Years in the Life of Our Planet*. Granta Books. Highly readable and nicely quirky account of the working of past (and future) plate tectonics.

Redfern, Martin. 2012. *The Earth: A Very Short Introduction*. Oxford University Press. Excellent and remarkably comprehensive on the inner workings of the Earth.

Rudwick, Martin. 2014. *Earth's Deep History*. University of Chicago Press. A scholarly though highly accessible introduction to how geology emerged as a science.

Stow, Dorrik. 2010. *Vanished Ocean: How Tethys Reshaped the World*. Oxford University Press. A story of the life and death of an ocean, and how this history shaped the landscapes around it.

Index

D

E

F

G